Identification of Promising Naval Aviation Science and Technology Opportunities

Committee on Identification of Promising Naval Aviation
Science and Technology Opportunities

Naval Studies Board

Division on Engineering and Physical Sciences

NATIONAL RESEARCH COUNCIL
OF THE NATIONAL ACADEMIES

THE NATIONAL ACADEMIES PRESS
Washington, D.C.
www.nap.edu

THE NATIONAL ACADEMIES PRESS 500 Fifth Street, N.W. Washington, DC 20001

NOTICE: The project that is the subject of this report was approved by the Governing Board of the National Research Council, whose members are drawn from the councils of the National Academy of Sciences, the National Academy of Engineering, and the Institute of Medicine. The members of the committee responsible for the report were chosen for their special competences and with regard for appropriate balance.

This study was supported by Contract No. N00014-00-G-0230, DO #25, between the National Academy of Sciences and the Department of the Navy. Any opinions, findings, conclusions, or recommendations expressed in this publication are those of the author(s) and do not necessarily reflect the views of the organizations or agencies that provided support for the project.

International Standard Book Number 0-309-09729-0

Copies of this report are available from:

Naval Studies Board, National Research Council, The Keck Center of the National Academies, 500 Fifth Street, N.W., Room WS904, Washington, DC 20001; and

The National Academies Press, 500 Fifth Street, N.W., Lockbox 285, Washington, DC 20055; (800) 624-6242 or (202) 334-3313 (in the Washington metropolitan area); Internet, http://www.nap.edu.

Printed in the United States of America

THE NATIONAL ACADEMIES

Advisers to the Nation on Science, Engineering, and Medicine

The **National Academy of Sciences** is a private, nonprofit, self-perpetuating society of distinguished scholars engaged in scientific and engineering research, dedicated to the furtherance of science and technology and to their use for the general welfare. Upon the authority of the charter granted to it by the Congress in 1863, the Academy has a mandate that requires it to advise the federal government on scientific and technical matters. Dr. Ralph J. Cicerone is president of the National Academy of Sciences.

The **National Academy of Engineering** was established in 1964, under the charter of the National Academy of Sciences, as a parallel organization of outstanding engineers. It is autonomous in its administration and in the selection of its members, sharing with the National Academy of Sciences the responsibility for advising the federal government. The National Academy of Engineering also sponsors engineering programs aimed at meeting national needs, encourages education and research, and recognizes the superior achievements of engineers. Dr. Wm. A. Wulf is president of the National Academy of Engineering.

The **Institute of Medicine** was established in 1970 by the National Academy of Sciences to secure the services of eminent members of appropriate professions in the examination of policy matters pertaining to the health of the public. The Institute acts under the responsibility given to the National Academy of Sciences by its congressional charter to be an adviser to the federal government and, upon its own initiative, to identify issues of medical care, research, and education. Dr. Harvey V. Fineberg is president of the Institute of Medicine.

The **National Research Council** was organized by the National Academy of Sciences in 1916 to associate the broad community of science and technology with the Academy's purposes of furthering knowledge and advising the federal government. Functioning in accordance with general policies determined by the Academy, the Council has become the principal operating agency of both the National Academy of Sciences and the National Academy of Engineering in providing services to the government, the public, and the scientific and engineering communities. The Council is administered jointly by both Academies and the Institute of Medicine. Dr. Ralph J. Cicerone and Dr. Wm. A. Wulf are chair and vice chair, respectively, of the National Research Council.

www.national-academies.org

COMMITTEE ON IDENTIFICATION OF PROMISING NAVAL AVIATION SCIENCE AND TECHNOLOGY OPPORTUNITIES

JOSEPH B. REAGAN, Saratoga, California, *Chair*
FRANK ALVIDREZ, Lockheed Martin Corporation
ALFRED O. AWANI, Boeing Company
WILLARD R. BOLTON, Sandia National Laboratories
WILLIAM C. BOWES, Morro Bay, California
H. LEE BUCHANAN, Perceptis, LLP
JOHN A. CORDER, Colleyville, Texas
ROBERT W. DAY, Raytheon Corporation
EARL H. DOWELL, Duke University
VALERIE J. GAWRON, General Dynamics
FRANK A. HORRIGAN, Bedford, Massachusetts
ARUN R. PALUSAMY, Northrop Grumman Electronic Systems
ROBERT J. POLUTCHKO, Charles Stark Draper Laboratory
BRUCE POWERS, George Washington University; Naval Postgraduate School
LYLE H. SCHWARTZ, Chevy Chase, Maryland
WILLIAM A. SIRIGNANO, University of California at Irvine

Staff

CHARLES F. DRAPER, Director
JAMES E. KILLIAN, Study Director
SUSAN G. CAMPBELL, Administrative Coordinator
MARY G. GORDON, Information Officer
IAN M. CAMERON, Research Associate
AYANNA N. VEST, Senior Program Assistant (as of June 25, 2005)
SIDNEY G. REED, JR., Consultant
RAYMOND S. WIDMAYER, Consultant

NAVAL STUDIES BOARD

Navy Liaison Representatives

Marine Corps Liaison Representative

LTGEN EDWARD HANLON, JR., USMC, Commanding General, Marine Corps Combat Development Command (through September 30, 2004)
LTGEN JAMES N. MATTIS, USMC, Commanding General, Marine Corps Combat Development Command (as of October 1, 2004)

Staff

CHARLES F. DRAPER, Director
ARUL MOZHI, Senior Program Officer
SUSAN G. CAMPBELL, Administrative Coordinator
MARY G. GORDON, Information Officer
IAN M. CAMERON, Research Associate
AYANNA N. VEST, Senior Program Assistant (as of June 25, 2005)

Preface

The Department of Defense (DOD) seeks to transform the nation's armed forces to meet the military challenges of the future. The Navy and the Marine Corps have defined their respective Service visions of transformation in Sea Power 21[1] and Marine Corps Strategy 21,[2] and together they form Naval Power 21,[3] the vision of how the naval forces of the United States will be equipped, trained, educated, organized, and employed in the 21st century. Joint Vision 2020[4] is the DOD vision that defines how the various elements of the DOD, including the naval forces, will operate in global conflicts as a single, integrated war-fighting entity. Many new war-fighting concepts are expressed in Naval Power 21, such as sea basing and network-centric operations, and the Office of Naval Research (ONR), in accordance with its mission to foster innovation in fields relevant to the Naval Services, requested that the National Research Council's Naval Studies Board conduct a study to identify new science and technology opportunities that might lead to new capabilities in naval aviation to support and enable these new war-fighting concepts.

---

[1]ADM Vern Clark, USN, Chief of Naval Operations. 2002. "Sea Power 21," *U.S. Naval Institute Proceedings*, Vol. 128, No. 10, pp. 32-41.

[2]Gen James L. Jones, USMC, Commandant of the Marine Corps. 1999. *Marine Corps Strategy 21*, Department of the Navy, Washington, D.C., July.

[3]Hon. Gordon England, Secretary of the Navy; ADM Vern Clark, USN, Chief of Naval Operations; and Gen James L. Jones, USMC, Commandant of the Marine Corps. 2002. *Naval Power 21 . . . A Naval Vision*, Department of the Navy, Washington, D.C., October.

[4]U.S. Joint Chiefs of Staff. 2000. *Joint Vision 2020*, Department of Defense, Washington, D.C.

The charge to the Committee on Identification of Promising Naval Aviation Science and Technology Opportunities specified related tasks that can be paraphrased as follows: (1) recognize what the Navy leadership has pronounced as the most important operational concepts of the future (e.g., the Naval Power 21 vision); (2) determine what capabilities are critical for implementing those operational concepts, especially as they apply to naval aviation; and (3) identify the technologies required to best enable those critical capabilities (i.e., assess the ONR science and technology (S&T) portfolio to enable capabilities and fill capability gaps). The terms of reference are given in full in Appendix A.

The complete process of constructing and implementing a research and development portfolio also includes other important tasks that were beyond the committee's charter of investigation: (4) an assessment of the state of maturity of each of the technologies to be developed, (5) prioritization of the work to be done and allocation of resources, and (6) the design of a transfer plan for the transition of each technology to a user. All of these tasks are, of course, interrelated, and they must be organized and prioritized in what is usually referred to as a strategic plan. The committee found no such plan at ONR to review. Although it did consider building a full naval aviation strategic S&T plan of its own to use as a template for its deliberations, the committee decided that such an activity was both well beyond its resources and would preempt the Navy's own process.

To illustrate the value of the strategic planning process, the committee first studied the concepts described in Naval Power 21 and considered the thoughts of some influential thinkers to gain insight into what these concepts imply for naval aviation; drew from its members' own experience and expertise to specifiy some capabilities that, if developed, would make a significant difference in naval aviation's future capabilities; and finally, sought to identify key technologies in which ONR could invest to achieve these capabilities. This report discusses the results of those efforts.

Acknowledgment of Reviewers

This report has been reviewed in draft form by individuals chosen for their diverse perspectives and technical expertise, in accordance with procedures approved by the National Research Council's Report Review Committee. The purpose of this independent review is to provide candid and critical comments that will assist the institution in making its published report as sound as possible and to ensure that the report meets institutional standards for objectivity, evidence, and responsiveness to the study charge. The review comments and draft manuscript remain confidential to protect the integrity of the deliberative process. We wish to thank the following individuals for their review of this report:

MajGen Charles F. Bolden, Jr., USMC (retired), Houston, Texas,
Eugene E. Covert, Massachusetts Institute of Technology,
Jose B. Cruz, Jr., Ohio State University,
Alan H. Epstein, Massachusetts Institute of Technology,
RADM Robert H. Gormley, USN (retired), The Oceanus Company,
James D. Lang, La Jolla, California,
Alton D. Romig, Jr., Sandia National Laboratories, and
Robert E. Whiteside, Henrico, North Carolina.

Although the reviewers listed above provided many constructive comments and suggestions, they were not asked to endorse the conclusions or recommendations, nor did they see the final draft of the report before its release. The review of this report was overseen by Lee M. Hunt of Alexandria, Virginia. Appointed by the National Research Council, he was responsible for making certain that an independent examination of this report was carried out in accordance with institutional procedures and that all review comments were carefully considered. Responsibility for the final content of this report rests entirely with the authoring committee and the institution.

Contents

APPENDIXES

Executive Summary

CONTEXT

The Committee on Identification of Promising Naval Aviation Science and Technology Opportunities was asked to examine the naval and joint operational concepts embraced in Sea Power 21 and Marine Corps Strategy 21, together known as Naval Power 21,[1] and to identify naval aviation capabilities that would enable these operational concepts. The committee was also asked to recommend science and technology (S&T) opportunities to the Office of Naval Research (ONR) that could support these future naval aviation capabilities and address any capability gaps.[2] However, the committee was not presented with any vision, strategy, or implementation plans by the Naval Air Systems Command (NAVAIR) or ONR regarding the role of naval aviation in satisfying the goals of Naval Power 21. Future capabilities that might be deployed were not identified, nor were existing capability gaps discussed at any length.

Based on its examination of the naval and joint operational concepts implicit in Naval Power 21 and drawing on the collective experience and expertise of its members, the committee identified seven "disruptive" capabilities inherent in or

---

[1]See ADM Vern Clark, USN, Chief of Naval Operations, 2002, "Sea Power 21," *U.S. Naval Institute Proceedings*, Vol. 128, No. 10, pp. 32-41; Gen James L. Jones, USMC, Commandant of the Marine Corps, 1999, *Marine Corps Strategy 21*, Department of the Navy, Washington, D.C., July; and Hon. Gordon England, Secretary of the Navy, ADM Vern Clark, USN, Chief of Naval Operations, and Gen James L. Jones, USMC, Commandant of the Marine Corps, 2002, *Naval Power 21 . . . A Naval Vision*, Department of the Navy, Washington, D.C., October.

[2]The terms of reference are given in full in Appendix A.

implied by Naval Power 21 (a list meant to be illustrative rather than exhaustive)—that is, capabilities that would profoundly change current modes of operation, greatly improve the effectiveness of war fighting, and contribute significantly to the realization of Naval Power 21.

Each of these capabilities—multispectral defense, unmanned air operations, hypersonic weapons delivery, fast-kill weapons, heavy-lift air transport, intelligent combat information management, and omniscient intelligence—is discussed in this report in terms of its benefits to naval aviation, why each is considered disruptive, and how each relates to at least one or more of the four pillars of Naval Power 21.

In addition, the committee addressed some of the S&T opportunities and focused development efforts required to make the disruptive capabilities a reality for naval aviation and Naval Power 21. As is the case for the capabilities list, the set of S&T opportunities discussed is not exhaustive. Committee members used their experience and expertise to provide a high-level assessment and to suggest where emphasis should be placed with respect to investments in Discovery and Invention (D&I) programs (6.1 and early 6.2) and Exploitation and Deployment (E&D) programs (late 6.2 and 6.3). The D&I programs tend to be longer term and higher risk and for the purposes of this study tend to fall into the 2011 to 2025 time frame. The shorter-term, generally less technically risky E&D programs are aimed for early insertion into the fleet and transition in the 2007 to 2010 time frame. Wherever possible, the committee categorizes the S&T opportunities as (1) naval unique (required only by naval missions), (2) naval essential (important for naval missions and non-naval missions), and (3) naval relevant (useful for both naval and non-naval missions).

Not intended as an in-depth technical review of the current naval aviation programs at ONR, this study identifies promising naval aviation S&T opportunities and capabilities that might enable, in the time frames indicated, the naval and joint operational concepts expressed in Naval Power 21, the Navy and Marine Corps strategic vision of future war fighting, and Joint Vision 2020.

STRATEGIC PLANNING

Naval aviation badly needs a clearly stated vision and strategic plan to focus its future. Moreover, NAVAIR and the Office of the Chief of Naval Operations (OPNAV) have the primary responsibilities for creating a naval aviation strategic S&T plan that identifies needed capabilities and the technology developments that can, over time, provide those capabilities. ONR (with the Naval Research Laboratory (NRL)) must be an essential partner with NAVAIR in developing a naval aviation strategic S&T plan.

During the course of this study, the Chief Technology Officer of NAVAIR acknowledged to the committee the need for such an S&T plan and agreed to develop one over the following year in conjunction with ONR. The Chief of

Naval Research addressed the committee and agreed that ONR would work closely with NAVAIR in the development of a strategic S&T plan for naval aviation. NAVAIR Program Executive Offices along with ONR/NRL scientists and technologists must be active participants in the creation of this plan. One of the goals should be to create a much closer strategic partnership between the organizations than currently exists.

S&T PLANNING AND EXECUTION

As part of its study, the committee was able to observe how S&T activities in naval aviation at ONR were organized, planned, funded, and executed and also was briefed on the planning processes used by the Army and the Air Force. The committee was thus able to compare the technical portfolio approach used at ONR and the systems enginnering approach, called integrated product and process development, used by the Air Force, and it developed recommendations on ways to improve the S&T planning and execution processes at ONR. The committee also learned that congressional add-ons constitute a significant fraction of the ONR funding for naval aviation S&T—a cause for concern since the funding for these projects supplants core S&T program funding and distorts strategic planning by inserting short-term, unanticipated projects that historically have not resulted in new capabilities for naval aviation. As ONR develops a strategic naval aviation S&T plan in response to the goals of Naval Power 21, the committee hopes that congressional add-ons will be replaced by core funding.

FINDINGS

Finding 1. NAVAIR currently lacks a naval aviation strategic plan that identifies capability gaps and technology development needs. A technology development plan established in cooperation with ONR does not exist. NAVAIR and ONR acknowledged the lack of a strategic plan for naval aviation's role in Naval Power 21. Both agreed that a naval aviation strategic S&T plan was essential, and both agreed to remedy the situation. As this study was being finalized, NAVAIR drafted the *Naval Aviation Vision 2020* document.[3] The committee believes this is a step in the right direction in forming the basis for such a strategic plan.

---

[3]As a result of a cooperative effort sparked by the present study, NAVAIR and ONR have issued the document *Naval Aviation Vision 2020* (see VADM James M. Zoortman, USN, Commander, Naval Air Forces; VADM Walter B. Massenburg, USN, Commander, Naval Air Systems Command; and RDML Thomas J. Kilcline, Jr., USN, Director, Air Warfare Division, 2005, *Naval Aviation Vision 2020*, Naval Aviation Enterprise, Department of the Navy, Washington, D.C. Available online at <http://www.nae.cnaf.navy.mil/demo/main.asp?ItemID=12>. Last accessed on September 30, 2005).

Finding 2. The concepts expressed in Naval Power 21 reflect a farsighted, aggressive, and challenging vision of future naval warfare for which neither a strategic operational plan nor a detailed implementation plan yet exists. Thus, capability needs and gaps for naval aviation have not yet been formally identified.

Finding 3. The strategic S&T planning processes of both the Army and the Air Force contain much that the committee believes could help the Navy in its planning process.

Finding 4. Current ONR planning appears to be largely ad hoc, with unclear goals against which to assess progress or ultimate value. The committee was unable to assess the relevance of current naval aviation S&T programs funded by ONR or their completeness in furnishing needed capabilities for Naval Power 21. No institutional process is currently in place at ONR to create or contribute to a vision of naval aviation for the future.

Finding 5. ONR's organization according to technical discipline makes it difficult for ONR to support cross-disciplinary areas, such as naval aviation. ONR currently lacks a formal process for managing naval aviation S&T, which involves multiple disciplines and programs located in six different ONR organizations. There is currently no single program manager with authority to approve a budget and long-term planning/direction setting for naval aviation S&T across ONR.

Finding 6. ONR does not use a systems engineering approach in the planning and execution of its technology development. As a result, projects are developed ad hoc and appear to be "opportunity" driven rather than "requirements" driven. Technology gaps are not systematically identified and thus are not well defined. Systems analysis is not used to determine technology priorities or investment strategies.

Finding 7. The committee believes that the large number of congressionally directed aviation projects at ONR is counterproductive to ONR's naval aviation S&T efforts. These projects supplant the budget for core S&T efforts, add to the workload of administrators and managers, and distort planning with the introduction of short-term, unexpected projects that rarely transition into future naval capabilities. The committee views current congressional add-ons not as a measure of success for ONR, but rather as a burden and a distortion of good S&T practice.

Additional findings are presented in Chapters 2 through 4.

RECOMMENDATIONS

Recommendation 1. To enable the capabilities for naval aviation operations as envisioned in Naval Power 21, the Chief of Naval Research, in partnership with NAVAIR, should lead the development of a naval aviation strategic S&T plan.

As this study was being finalized, the Commander of Naval Air Forces, the Commander of NAVAIR, and the Director of the Air Warfare Division in OPNAV created the document *Naval Aviation Vision 2020*, which can provide a basis for the development of this strategic S&T plan. This plan should be updated annually in synchronization with the Planning, Programming, Budgeting, and Execution System process. It should also be considered in the naval aviation S&T plans of the Army and the Air Force.

Recommendation 2. ONR should establish a formal process for the identification of key S&T approaches that will identify and address naval aviation capability gaps. A methodology should be developed for analyzing options and selecting preferred approaches based on a systems perspective that includes technology trade-offs, maturity, risks, cost, impact, and so on. A methodology should also be developed for connecting novel concepts and potential breakthroughs into a naval aviation strategic S&T plan such that they receive attention, undergo development, and have a path into the acquisition domain.

Recommendation 3. ONR should consider the S&T planning processes used by the Army and the Air Force as a source of potential guidance in developing a naval aviation strategic S&T plan.

Recommendation 4. As ONR develops a naval aviation strategic S&T plan, consideration should be given to the following disruptive aviation capabilities, each of which can be traced to at least one of the four components—Sea Shield, Sea Strike, Sea Basing, and FORCEnet—of Naval Power 21:

- Multispectral defense,
- Unmannned air operations,
- Hypersonic weapons delivery,
- Fast-kill weapons,
- Heavy-lift air transport,
- Intelligent combat information management, and
- Omniscient intelligence.

Science and technologies in which ONR could pursue advances to enable each of these capabilities are discussed in Chapter 3.

Recommendation 5. The Chief of Naval Research should establish a single point of responsibility for the development of a naval aviation strategic S&T plan at ONR.

This responsibility must include both budget and direction-setting authority, even though the technology development will occur in several different organizations. This would enable development of a prioritized, balanced, and well-integrated program that has a high probability of transitioning technology into the operational naval forces.

Recommendation 6. The Chief of Naval Research should strengthen ONR's analytic capabilities. A cadre of systems analysis personnel who can interface between the mission capability analysis personnel at OPNAV and the Navy Warfare Development Command and the scientists and technologists at ONR is needed to support strategic planning for naval aviation S&T.

Recommendation 7. With the establishment of a naval aviation strategic S&T plan and the identification of critical gaps in capabilities for naval aviation, ONR should inform and educate congressional staffers about technologies and capabilities that would significantly advance the closure of such gaps, thus turning a currently burdensome relationship into a strategic supportive force.

Additional recommendations are offered in Chapters 2 through 4.

1

Introduction

BACKGROUND

The mission of the Office of Naval Research (ONR) is to sponsor research and development (R&D) in fields relevant to the needs of the Navy and the Marine Corps (the Naval Services). ONR maintains relationships with the R&D communities in universities, industry, other government agencies (including the other Service branches), and with the operational communities in the Naval Services to understand their science and technology (S&T) needs, and it provides funding and manages S&T development activities across a broad range of disciplines by contracting with external groups that perform the research. For example, ONR provides base funding for the Naval Research Laboratory (NRL), a full-service R&D facility with laboratories, test chambers, test ranges, and a large scientific staff. ONR does not maintain its own laboratories or research facilities and does not have a large staff of scientists and engineers.

ONR investments in naval aviation support the R&D of manned and unmanned aircraft for offensive and defensive counterair operations/attack, strategic attack, interdiction, control of the sea lanes (including antisubmarine warfare), surveillance and reconnaissance, air support for ground troops, and air logistics. The principal interface with, and user of, the naval aviation S&T sponsored by ONR is the Naval Air Systems Command (NAVAIR). ONR's two programs for executing these investments are (1) Discovery and Invention (D&I), supporting longer-term, higher-risk basic and applied research efforts (categories 6.1 and early 6.2), and (2) Exploitation and Deployment (E&D), consisting of technology development and demonstration efforts (categories late 6.2 and 6.3) that tend to be shorter term, have reduced technological risk, and are aimed for early transition from S&T and insertion into the fleet. Current D&I efforts include naval-aviation-unique aircraft technology developments such as ship airwake modeling, fixed-wing composite structure corrosion fatigue analysis, and flight safety and autonomous control technologies in carrier operations. Future activities will include persistent aerial intelligence, surveillance, and reconnaissance capabilities using unmanned aerial vehicles (UAVs) targeted for expeditionary strike groups, as well as structurally embedded antennas, sensors, and avionics to be integrated with future airframes. In addition, ONR will work closely with the Air Force Sensor Craft program to leverage its extensive investment and development.[1]

Centered on Future Naval Capabilities (FNCs; see Appendix D), current E&D efforts are highly focused and managed by integrated product development teams (IPDTs) with the goal of achieving rapid transition of the resulting technology to the fleet. They include (1) exploration of UAV propulsion technologies and development of UAV intelligent autonomy as part of the Autonomous Operations FNC; (2) leveraging of investments in the Joint Unmanned Combat Air Vehicle program (sponsored by the Defense Advanced Research Projects Agency (DARPA) and the Air Force), part of the Time Critical Strike FNC; and (3) investment in the development of next-generation aircraft (manned) and cruise missile (expendable) turbine engine propulsion, part of the Total Ownership Cost FNC. ONR investments in aviation also include Marine Corps programs for heavy-lift rotorcraft and the reconfigurable rotor blade, as well as congressionally directed aviation investments, including the variable exhaust nozzle, DP-2, anticorrosion modeling software, integrated processor fuel cell, integrated aircraft health management, advanced thin-film coatings, and aviation ground navigation systems. ONR manages a number of congressionally mandated and funded aviation S&T programs that are not of its own selection.

The naval aviation S&T activities funded by ONR are not concentrated in a single organization but rather are conducted under the purview of several departments, referred to as codes. Furthermore, ONR is not organizationally structured according to war-fighting functional areas, such as naval aviation, surface ship warfare, and weapons systems, but instead along technical discipline lines, such as electronics, materials, and human systems. While such a structure is not uncommon in S&T organizations, it is often complemented by a strong program office structure representing, for example, war-fighting discipline areas led by individuals who are responsible for the funding and management of technology development across many organizational and technical disciplines. Without such

[1]The Air Force has the largest Department of Defense (DOD) investment in all S&T for fixed-wing vehicles (both manned and unmanned). The Army has the lead on all rotary-wing vehicles (both manned and unmanned) for the DOD. ONR is planning to work closely with both the Air Force and the Army to leverage their much larger programs and avoid duplication so as to enable the Navy to pursue naval-unique applications.

a strong focus it is difficult to achieve an integrated and efficient technical program that has a high probability of developing technologies that will transfer to the fleet. The IPDT-managed FNCs focus on the transfer of existing high-priority capabilities such as time-critical strike, and not on a broader war-fighting functional capability, such as naval aviation.

The largest aviation technology development activity at ONR resides in one of ONR's six main departments, Code 35, Naval Expeditionary Warfare, but significant development in sensors, information, and electronics for aircraft is conducted in Code 31, in materials for aircraft in Code 33, and to a lesser extent in ocean science and human systems in Codes 32 and 34, respectively. There is no single program manager at ONR responsible for the entire funding, management, and technical direction of S&T activities as they relate to naval aviation. Details on the naval aviation program at ONR, including its organization, program structure, and funding allocations, are given in Appendix D.

As one means of ensuring that its investments appropriately address naval priorities and requirements and that its programs are of high scientific and technical quality, ONR senior management requires that each of its departments undergo a review every 3 years. Several such reviews have been conducted on various programs within Code 35 over the past several years by various committees of the Naval Studies Board (NSB) of the National Research Council (NRC).

In its 1999 *Assessment of the Office of Naval Research's Air and Surface Weapons Technology Program,*[2] there was concern that project selection was methodological rather than strategic, that the S&T work was evolutionary in nature and focused on short-term needs, and that trade-off studies needed to be conducted to determine how to fit the 6.2 and 6.3 program components into the overall weapons system architecture.

In its 2001 *Assessment of the Office of Naval Research's Aircraft Technology Program,*[3] there was concern that the technical program was not influenced by a long-range vision or strategic planning for the future of naval aircraft technology. To that end, the report recommended that the staff of the Office of the Chief of Naval Operations (OPNAV), in conjunction with NAVAIR and the ONR, develop a long-range naval aviation strategic plan that would include an S&T plan. It was further stated that such planning should provide (1) a framework for future ONR S&T investments, including significant emphasis on D&I, and (2) a vision for new capabilities, including advanced air vehicle concepts at affordable costs.

---

[2]Naval Studies Board, National Research Council. 1999. *1999 Assessment of the Office of Naval Research's Air and Surface Weapons Technology Program*, National Academy Press, Washington, D.C.

[3]Naval Studies Board, National Research Council. 2001. *2001 Assessment of the Office of Naval Research's Aircraft Technology Program*, National Academy Press, Washington, D.C.

In its 2002 *Assessment of the Office of Naval Research's Air and Surface Weapons Technology Program,*[4] it was recommended that, in collaboration with other Department of the Navy elements, ONR should develop a strategic naval air and surface weapons technology plan that would achieve a balance between near- and long-term goals. Moreover, the use of systems analysis both in developing the strategic plan and in S&T planning overall at ONR was needed.

THIS STUDY

At the request of ONR, the NRC, under the auspices of the NSB, established the Committee on Identification of Promising Naval Aviation Science and Technology Opportunities in September 2004 to identify promising naval aviation S&T opportunities in basic research (6.1), applied research (6.2), and advanced technology development (6.3) areas. The full terms of reference are given in Appendix A.

Not intended to be an in-depth technical review of the current naval aviation programs at ONR, the current study focuses on identifying promising naval aviation S&T opportunities and capabilities that might enable the naval and joint operational concepts expressed in Naval Power 21[5] and Joint Vision 2020.[6] The Navy and Marine Corps have defined their respective Service visions in Sea Power 21[7] and Marine Corps Strategy 21,[8] and together they form Naval Power 21,[9] the vision of how the naval forces of the United States will be equipped, trained, educated, organized, and employed in the 21st century. Joint Vision 2020[10] is the Department of Defense (DOD) vision that defines how the various elements of the DOD, including the naval forces, will operate in global conflicts as a single, integrated, war-fighting entity. There are many new war-fighting concepts expressed in Naval Power 21,[11] such as sea basing and network-centric

---

[4]Naval Studies Board, National Research Council. 2002. *2002 Assessment of the Office of Naval Research's Air and Surface Weapons Technology Program,* National Academy Press, Washington, D.C.

[5]Hon. Gordon England, Secretary of the Navy; ADM Vern Clark, USN, Chief of Naval Operations; and Gen James L. Jones, USMC, Commandant of the Marine Corps. 2002. *Naval Power 21 . . . A Naval Vision,* Department of the Navy, Washington, D.C., October.

[6]U.S. Joint Chiefs of Staff. 2000. *Joint Vision 2020,* Department of Defense, Washington, D.C.

[7]ADM Vern Clark, USN, Chief of Naval Operations. 2002. "Sea Power 21," *U.S. Naval Institute Proceedings*, Vol. 128, No. 10, pp. 32-41.

[8]Gen James L. Jones, USMC, Commandant of the Marine Corps. 1999. *Marine Corps Strategy 21,* Department of the Navy, Washington, D.C., July.

[9]Hon. Gordon England, Secretary of the Navy; ADM Vern Clark, USN, Chief of Naval Operations; and Gen James L. Jones, USMC, Commandant of the Marine Corps. 2002. *Naval Power 21 . . . A Naval Vision,* Department of the Navy, Washington, D.C., October.

[10]U.S. Joint Chiefs of Staff. 2000. *Joint Vision 2020,* Department of Defense, Washington, D.C.

[11]Hon. Gordon England, Secretary of the Navy; ADM Vern Clark, USN, Chief of Naval Operations; and Gen James L. Jones, USMC, Commandant of the Marine Corps. 2002. *Naval Power 21 . . . A Naval Vision,* Department of the Navy, Washington, D.C., October.

operations, and ONR requested in this study that the committee identify new S&T opportunities and capabilities in naval aviation that would enable those concepts.

The committee held two meetings in Washington, D.C., on September 28-30, 2004, and on October 26-27, 2004, to gather information about Naval Power 21, ONR, the naval aviation S&T program at ONR across all of the relevant organizations, and the aviation technology program activities at NAVAIR and within the Marine Corps. The committee did not receive a comprehensive briefing on each of the aviation programs and projects within ONR since a detailed technical review was not the goal of the study. The committee heard presentations from the Chief of Naval Research, the Technical Director of ONR, the Chief Technology Officer of NAVAIR, the Deputy Assistant Commander for Research and Engineering at NAVAIR, the S&T Director of the Air Warfare Division of OPNAV, the Deputy Commandant for Aviation, Headquarters Marine Corps, the Director of Research at NRL, the Director of Weapons Systems in the Office of the Secretary of Defense Research and Engineering, the Sea Trial Director of the Navy Warfare Development Command, and the Deputy Assistant Secretary of the Navy for Research, Development, Test, and Evaluation (RDT&E).

The committee also received briefings from the Army Aviation Applied Technology Directorate, the Air Vehicles Directorate at the Air Force Research Laboratory, and the Office of Aerospace Technology at the National Aeronautics and Space Administration (NASA). From these presentations the committee learned a great deal about the current S&T planning processes within the Army, the Air Force, and NASA. A study conducted by the Deputy Assistant Secretary of the Navy[12] provided the committee with considerable information about the overall balance and relevance of the S&T programs and planning processes at ONR. The committee also heard from the Tactical Technology Office at DARPA.

The committee held a final meeting at the National Academies' facility in Irvine, California, on November 16-18, 2004, to reach consensus on the findings and recommendations and to prepare a first draft of the final report.

EMPHASIS AND APPROACH IN THIS STUDY

The committee emphasizes the critical importance of good strategic planning to the success of any enterprise. It is only through the creation of a vision of a desired future state—the development of a strategic plan for achieving that vision followed by a detailed tactical implementation plan—that the ultimate goal can be achieved. The committee believes that (1) good strategic planning is critical to the success of naval aviation and (2) naval aviation is critical to the success of Sea

[12]Michael McGrath, Deputy Assistant Secretary of the Navy (RDT&E), "DoN S&T Planning," presentation to the committee on October 26, 2004.

Power 21 and Marine Corps Strategy 21 (together constituting Naval Power 21). Good technology planning and capability development can be achieved only within the context of a larger strategic plan. Without a strategic plan, aviation S&T projects will be merely a collection of unrelated activities.

Chapter 2 of this report discusses the committee's findings regarding the strategic planning processes for naval aviation and specifically for S&T activities at ONR. At the beginning of the study, a naval aviation strategic plan did not exist at NAVAIR or OPNAV. There was neither a master plan for the role that naval aviation would play in Naval Power 21, nor a strategic aviation S&T plan to guide the efforts of NAVAIR and ONR. The committee observed that the interactions between these two key organizations were not strategic but instead opportunistic, despite the numerous recommendations made by previous NRC studies that the Navy develop a strategic S&T plan. Due to the lack of a coherent S&T planning process, technology gaps and desired capabilities related to Naval Power 21 were not identified to the committee.

As the study progressed, the committee received verbal commitments from the Chief Technology Officer of NAVAIR, representing the Commander of NAVAIR, that NAVAIR would develop a vision and a naval aviation strategic S&T plan that would address the goals of Naval Power 21. The Chief of Naval Research agreed to work with NAVAIR to develop this plan. As this study was being finalized, the committee learned that such a road map for naval aviation in the 21st century, *Naval Aviation Vision 2020*,[13] has now been created by the Commander of Naval Air Forces, Director, Air Warfare Division in OPNAV, and the Commander of NAVAIR.

Lacking for use in its study a naval aviation strategic plan at NAVAIR and OPNAV and a naval aviation strategic S&T plan at ONR for meeting the requirements of Naval Power 21, the committee developed the following approach to recommending future capabilities and associated technology developments that could form the basis of a revitalized aviation S&T program at ONR. Based on their examination of the visionary concepts and context expressed in Naval Power 21 and their personal experiences and knowledge of how the desired goals might be met, the committee's members chose a set of of "disruptive" capabilities that, if developed by naval aviation, would have a profound effect on future network-centric warfare. The set of seven capabilities described in Chapter 2 was selected from a much larger suite of desirable capabilities considered by the committee

[13]VADM James M. Zortman, USN, Commander, Naval Air Forces; VADM Walter B. Massenburg, USN, Commander, Naval Air Systems Command; and RDML Thomas J. Kilcline, Jr., USN, Director, Air Warfare Division. 2005. *Naval Aviation Vision 2020,* Department of the Navy, Washington, D.C. Available online at <http://www.nae.cnaf.navy.mil/demo.main.asp?ItemID=12>. Last accessed on September 30, 2005.

and is not meant to be exhaustive. As ONR develops its naval aviation strategic S&T plan, the committee is confident that these capabilities will rank high in consideration.

In Chapter 3 the committee identifies the technology opportunities and developments that would be necessary to make the "disruptive" capabilities a reality. Committee members used their experience and expertise in relevant areas to provide a high-level assessment of technology options and to suggest where emphasis should be placed. The committee has attempted to categorize these technology developments as naval unique, naval essential, or naval relevant and to position them in the 2007 to 2010 and 2011 to 2025 time frames wherever possible. These technology development areas can provide the basis of a revitalized naval aviation S&T program at ONR, and will go a long way toward positioning naval aviation as a major contributor to the success of Naval Power 21.

Finally, because of the importance of strategic planning to the success of any enterprise, the committee in Chapter 4 addresses the S&T planning processes used at ONR and makes some recommendations for change based on the successful practices of other Service branches. The committee hopes that these recommendations will be useful to ONR as it embarks on the development of a strategic S&T plan for naval aviation.

The terms of reference, short biographies of committee members, and information on study activities are given in Appendixes A through C, respectively. Allocation of funding for ONR's naval aviation program is discussed in Appendix D. Acronyms used in the report are defined in Appendix E.

2

Strategic and Technology Planning and Disruptive Capabilities for Naval Aviation

RATIONALE

The committee received a number of briefings and reviewed several documents to gain insight into naval aviation's role in enabling the operational concepts envisioned for the future in Naval Power 21 and to assess the current situation at ONR with respect to the planning, funding, and development of technology to support this thrust. Based on the material it reviewed, the committee was unanimous in its conviction that naval aviation plays a crucial role in the transformation to Naval Power 21. ONR's role in developing the necessary naval aviation technology for the future is thus extremely important to the success of Naval Power 21.

The committee received much information indicating that ONR has many S&T programs under way that are aimed at important aviation issues. Few of those programs were specific to naval aviation, although the aggregated funding that is relevant to aviation is a considerable fraction of the budget. Unfortunately, however, the committee found little that could be construed as a strategic plan to guide allocation of that funding to strengthen naval aviation. Investments appeared ad hoc, with unclear goals against which to assess progress or ultimate value. This perception of disarray was exacerbated by the very large fraction of programs funded by congressional largess (discussed in Chapter 4) that have no connection to Navy objectives and important operational concepts. Futhermore, the committee found that funding for basic research on the longest-range visions (6.1 funding) for naval aviation was insignificant, suggesting that ONR's current portfolio is not very forward-looking. As a result, the committee was unable to assess the relative relevance of ONR's current aviation programs or their completeness in furnishing the capabilities needed for realizing Naval Power 21.

This deficit was immediately acknowledged by both ONR and NAVAIR. Both agreed that an S&T plan, jointly crafted, is essential.[1] The committee did review worthy strategic plans of both the Army and the Air Force in this sector and found much (discussed further in Chapter 4) that could help ONR and NAVAIR in the essential task of developing a jointly crafted S&T plan.

Construction of a naval aviation strategic S&T plan will require obtaining input from the widest spectrum of users, producers, and technologists. For the leadership of ONR, the real value of the process is as much the intuition that is built in by the participants as the specifics of what is recommended. What is offered in this report is only illustrative of both the process and the possible outcome.

NAVAL POWER 21

Sea Power 21[2] and Marine Corps Strategy 21,[3] jointly referred to as Naval Power 21,[4] offer a farsighted, aggressive, challenging vision of future naval warfare, well beyond the reach of current U.S. naval capabilities. Needed future capabilities to achieve that vision are detailed in the discussions of the fundamental components of Sea Power 21 that follow.

Sea Power 21 identifies three fundamental concepts that will provide the foundation for the Navy's future effectiveness: Sea Strike, Sea Shield, and Sea Basing. Respectively, they enhance the U.S. ability to project offensive power, to provide defensive assurance, and to enhance operational independence around the globe.

Sea Strike is a broadened concept for naval power projection that leverages enhanced command, control, communications, computers, intelligence, surveillance, and reconnaissance (C4ISR), precision delivery of weapons, stealth of

---

[1]As a result of a cooperative effort sparked by the present study, NAVAIR and ONR have issued the document *Naval Aviation Vision 2020* (see VADM James M. Zoortman, USN, Commander, Naval Air Forces; VADM Walter B. Massenburg, USN, Commander, Naval Air Systems Command; and RDML Thomas J. Kilcline, Jr., USN, Director, Air Warfare Division, 2005, *Naval Aviation Vision 2020*, Naval Aviation Enterprise, Department of the Navy, Washington, D.C. Available online at <http://www.nae.cnaf.navy.mil/demo/main.asp?ItemID=12>. Last accessed on September 30, 2005).

[2]ADM Vern Clark, USN. 2002. "Sea Power 21: Projecting Decisive Joint Capabilities," *U.S. Naval Institute Proceedings*, Vol. 128, No. 10, pp. 32-41.

[3]Gen James L. Jones, USMC, Comandant of the Marine Corps. 2000. *Marine Corps Strategy 21*, U.S. Government Printing Office, Washington, D.C., November.

[4]Hon. Gordon England, Secretary of the Navy; ADM Vern Clark, USN, Chief of Naval Operations; and Gen James L. Jones, USMC, Commandant of the Marine Corps. 2002. *Naval Power 21 . . . A Naval Vision*, Department of Defense, Washington, D.C., October.

operation, and much extended persistence for increasing operational tempo, reach, and effectiveness. Explicitly included in the definition of Sea Strike are a number of potentially disruptive capabilities: for example, persistent intelligence, surveillance, and reconnaissance supported by autonomous (i.e., unmanned aerial vehicles, or UAVs) long-dwell sensors, covert strike, knowledge enhancement systems, unmanned combat air vehicles, hypersonic missiles, electromagnetic rail guns, and Ship-to-Objective Maneuver (STOM),[5] the centerpiece of the Marine Corps's capstone Expeditionary Maneuver Warfare concept, another disruptive capability with severe logistics, communications, and firepower support challenges.

Sea Shield, similarly, is a concept for broadened defense—taking the naval forces beyond the traditional defense of the unit and task force to provide the nation with sea-based theater and strategic defense. Incorporated in the Sea Shield concept are a number of potentially disruptive capabilities, including directed-energy weapons, UAVs, the single integrated air picture, and distributed weapons coordination.

Sea Basing, in turn, projects the sovereignty of the United States globally, providing Joint Force commanders with vital command and control, fire support, and logistics from the sea while minimizing reliance on vulnerable support assets and infrastructure ashore. While Sea Strike and Sea Shield are broadened versions of traditional naval functions, Sea Basing introduces a whole new concept of operating that in itself is clearly disruptive. Included in the description of the Sea Basing concept is explicit reference to individual disruptive capabilities, such as heavy equipment (at-sea) transfer, improved vertical delivery methods, and rotational crewing infrastructure.

There is an important fourth component of Sea Power 21—FORCEnet[6]—which ties the three primary components together. FORCEnet represents the information architectural framework and operational concept that interconnect all of the critical elements of the naval warfare enterprise into a seamless networked distributed combat force. Sea Power 21 defines FORCEnet explicitly as "the Navy's plan to make (network-centric warfare) an operational reality," including such concepts as "sensor and weapon grids," "distributed, collaborative C^2 [command and control]," "dynamic survivable networks," and "adaptive/automated decision aids."

The visionary concepts expressed in Sea Power 21 must be embodied in a strategic plan for naval aviation that could ultimately lead to a detailed imple-

[5]STOM envisions the Marines moving directly from a ship base to an objective, perhaps hundreds of miles inland, without setting up any secondary supporting deployments (e.g., beach depots, communication relays). Obviously, sea-based logistics is a necessity.

[6]ADM Vern Clark, USN, Chief of Naval Operations, and Gen Michael W. Hagee, USMC, Commandant of the Marine Corps. 2005. "FORCEnet: A Functional Concept for the 21st Century," Department of the Navy, Washington, D.C., February.

mentation plan. Such a strategic plan and a comprehensive implementation plan did not exist at the outset of this study.[7] For example, there are multiple references in Sea Power 21 to the use of UAVs, but the slow pace at which the Navy is adopting this capability into the fleet belies the vision. The Navy is in the process of designing and implementing elements of the infrastructure that will provide the underpinnings of Sea Power 21 and may exploit disruptive technologies, but it has done little to identify related concepts of operation. As a result, there is currently no strategy for how naval aviation will fight in the network-centric world. There have been no war games exploiting network-centric naval aviation operations. The Navy must explore, including joint developments with the other Services, the various options for operation in a network-centric world that deal with the uncertainties of world politics and the disruptive capabilities created by adversaries.

The Navy should trade between options for dealing with a wide range of contingencies and should develop a clear strategy for how it intends to operate in the future. Only then can an implementation plan and a detailed design for a successful system be undertaken by which naval air power can determine how it will fight in the future and, in particular, how it will exploit disruptive capabilities and technologies to win against any adversary that will develop its own disruptive capabilities. The Strategic Studies Group XXIII, which reports directly to the Chief of Naval Operations and is tasked with generating revolutionary concepts for future naval war fighting, came to a similar conclusion in its December 2004 report entitled *Beyond Maritime Supremacy—Transforming Maritime Forces for the Global Fight Against Terrorism,* recommending changes in policy, processes, organizational structure, resource allocation, and doctrine to implement the necessary actions.[8]

SOME DISRUPTIVE CAPABILITIES

Based on its understanding of the challenges implicit or explicit in Naval Power 21, the committee selected a subset of seven disruptive capabilities that seem particularly important for future naval aviation:

- Multispectral defense,
- Unmanned air operations,
- Hypersonic weapons delivery,
- Fast-kill weapons,
- Heavy-lift air transport,

---

[7]See footnote 1 in this chapter.

[8]ADM James R. Hogg, USN (Ret.), Director, CNO Strategic Studies Group, personal communication, August 10, 2005.

TABLE 2.1 Mapping of Seven Disruptive Capabilities to the Four Fundamental Concepts in Naval Power 21

Disruptive Capabilities	Sea Strike	Sea Shield	Sea Basing	FORCEnet
Multispectral defense	X	x		
Unmanned air operations	X	X	X	
Hypersonic weapons delivery	X	x		
Fast-kill weapons	X	X		
Heavy-lift air transport	x		X	
Intelligent combat information management	x	x		X
Omniscient intelligence	x	x		X

NOTE: A capital "X" indicates a direct reference to the capability in Naval Power 21; a lower-case "x" indicates a particularly relevant role as judged by the committee for the capability, although it may not have been explicitly identified as such in Naval Power 21.

- Intelligent combat information management, and
- Omniscient intelligence.

This list is not exhaustive but does illustrate the areas in which a comprehensive and coherent S&T program should be investing. Each area is discussed briefly below with an indication of the benefits to Sea Power 21, the S&T challenges, and the reasons that each is considered disruptive. Each can be traced directly to at least one of the four components of Naval Power 21—Sea Strike, Sea Shield, Sea Basing, and FORCEnet. Table 2.1 summarizes these relationships. FORCEnet, in particular, infuses everything.

Multispectral Defense

Covert strike is fundamental to the concept of Sea Strike. Naval aircraft will be expected to operate in environments that contain many different types of threats ranging from vintage radar-directed antiaircraft artillery and infrared heat-seeking missiles to the most sophisticated multimodal, hypersonic antiaircraft missiles. Simultaneous defense against all of these threats at a size and in a space that do not infringe on envelopes of operation is very challenging. Technologies and techniques that enable electronic warfare and stealth have for many years strengthened the Navy's capability for providing multispectral defense.

The use of stealth in the Gulf War made U.S. air power virtually immune to air defenses. As a result the United States destroyed most of the Iraqi communications, command and control, aircraft on the ground, and tanks, and even conducted daytime bombing. This change in operations was truly profound.

Regrettably, however, the growing capability and diversity of the threat faced by the United States require renewed emphasis beyond the traditional incremental approaches. Enemy air defenses are improving in their ability to counter the U.S. advantage of stealth. The need to fly low to support U.S. troops on the ground is making infrared and visual signatures very important. New technologies are needed to reduce these signatures.

Active and precise hard-kill countermeasures (e.g., kilowatt-class lasers) are now required to handle attackers in multiple regions of the spectrum. In both the optical and the microwave, for example, technologies are needed that will provide signatures similar to or capable of blending with the background. These technologies will involve new materials, tactics, and electronics common to all air platforms (winged aircraft, helicopters, short takeoff and vertical landing (STOVL) vehicles, and UAVs). The most disruptive of these technologies would be a visual stealth system that is agile, responds quickly, and is small enough and inexpensive enough to effectively protect low-altitude air vehicles (UAVs, helicopters, STOVL aircraft, and even fixed-wing aircraft).

Unmanned Air Operations

Unmanned air operations are, and will continue to become, increasingly important to naval aviation; they are explicitly identified by Sea Power 21 as a capability required for both Sea Strike and Sea Shield. It is envisioned that long-duration (days and weeks) loitering vehicles will perform missions such as surveillance with all manner of sensors; electronic attack; and communication relays with controllable and selectable/adaptive communications bandwidth. When time is of the essence, hypersonic (Mach 3 to 10) UAVs will get weapons and sensors to targets before the enemy can move very far.

UAVs currently provide reconnaissance, intelligence gathering, and surveillance with flight vehicles of various sizes serving in a variety of roles. Small, back-packable, hand-launched UAVs can provide a unit on maneuvers with images of its immediate surroundings; intermediate-sized UAVs suitable for transport in a single vehicle for forward basing typically carry electrooptical sensors to provide reconnaissance in some depth, or suitable for ship basing to provide area surveillance; and large, high-altitude, long-endurance UAVs with extensive sensor payloads provide broad surveillance across an area of operations.

In the future, UAVs will take on a larger range of roles, including combat and support missions currently performed by manned aircraft. The development begun with a previous centerpiece program (Unmanned Combat Aerial Vehicle-Navy) and the Joint Unmanned Combat Aerial System may result in high-performance, possibly hypersonic, armed UAVs that can perform tactical missions.

Among the many motivations for the aggressive introduction of UAVs into naval aviation are a reduction in the number of people placed in harm's way; the potentially superior performance of systems that do not have to accommodate a human occupant or can operate in environments not suitable for humans; an expansion of the size range of militarily useful systems not limited by the scale of a human occupant and facilitated by the development of miniaturized sensors and systems; and the potential for reduced life-cycle costs through reduced staffing.

It is vital that the Navy embrace the rapidly developing UAV technologies and prepare for their wider introduction into the fleet. In 2000, the National Research Council's Committee on Uninhabited Air Vehicles identified opportunities for research on crosscutting UAV subsystem technologies.[9]

To realize the vision of Naval Power 21, UAVs must be affordable; must have sufficient range, endurance, and payload to meet the missions of Sea Strike and Sea Shield; and must have highly integrated airframes, propulsion systems, sensors, communication, self-protection, and weapons. They must be completely integrated into the networked war-fighting system (FORCEnet) as an integral part of the envisioned network-centric sensor grids or possibly as a portion of a distributed sensor system composed of a number of cooperating UAV platforms.

It is essential that the Navy foster the development of UAVs that are suitable for its unique operational environments and missions through a well-planned process of S&T development that will support the required capabilities in a timely manner to maintain technological and operational superiority in the future. Although much work has been done to develop the potential of UAVs, there are aspects of their application in naval operations that are unique and should be the subject of focused Navy research and development. For example, naval UAVs will operate in a maritime environment with its unique humidity, exposure to saltwater, and extremes of temperature; they will operate from ships in all weather conditions with inherent platform motion; and they will have to take off, land, and maneuver in close proximity to other aircraft, both remotely operated and occupied. In addition, there are substantial planning and operational challenges associated with operations from ships, which frequently provide limited alternate landing sites and have severe limitations on landing opportunities.

For Marines the operation of backpack-size, small UAVs in conjunction with tactical units ashore and in severe environments, with minimal logistics and less than delicate handling, offers additional challenges. Nevertheless, second only to the impact of network-centric operations, the widespread application of UAV technologies will enormously alter the way the Navy goes about its everyday business.

Hypersonic Weapons Delivery[10]

The Sea Strike component of Sea Power 21 includes the timely delivery of ordnance to satisfy the military objectives of projecting power, supporting ground operations, and attacking time-critical targets. To minimize the time from identi-

---

[9]National Research Council. 2000. *Uninhabited Air Vehicles: Enabling Science for Military Systems*, National Academy Press, Washington, D.C.

[10]For a longer discussion of this topic, see National Research Council, 2004, *Evaluation of the National Aerospace Initiative*, The National Academies Press, Washington, D.C.

fication and fixing of a target until strike, delivery systems may be forward based, loitering in the target area, or capable of high speed to minimize time to target. In-depth studies are needed to evaluate the systems-level cost and performance trade-offs among the options. However, one technology area that is likely to have high value is hypersonic (e.g., Mach 3 to 10) flight.[11]

Currently, the U.S. ability to attack mobile targets is still compromised because of the lack of persistent surveillance and the inability to instantly execute a kill. Often opposing mobile missile launchers move to take cover in civilian areas that U.S. forces cannot attack by the time they have a weapon ready for delivery. More continuous surveillance (e.g., via multiple UAVs) coupled with hypersonic kill vehicles could yield a near-instantaneous response, giving an enemy little opportunity to hide. Such a capability would profoundly change the nature of warfare as we know it and would be truly disruptive.

Air-breathing hypersonic vehicles obtain thrust through a portion of their flight trajectory by burning a fuel with oxygen obtained from the atmosphere, thus gaining a weight advantage by avoiding having to carry an oxidizer and the systems to contain and control it. Air breathing implies operation in the atmosphere during a vehicle's propulsive flight phase, which presents combustion system and thermal management challenges.

Self-contained propulsion systems carry both fuel and oxidizer, resulting in a weight penalty, but have the advantage that their operation is based on well-established rocket technology. A self-contained propulsion system can operate both in the atmosphere and outside it.

Both types of propulsion system could be used to provide continuous thrust during a significant portion of a vehicle's flight trajectory or to boost the vehicle to high speed, with the kinetic energy used during the remainder of the trajectory for range extension (assuming that the hypersonic vehicles of interest have their own navigation, guidance, and control systems rather than follow a ballistic trajectory). With either type of propulsion system, hypersonic vehicles would have in common a capacity for high speed (potentially resulting in a valuable reduction in time to target). They would also pose technical challenges such as thermal management, durability of materials, control systems, sensing and communications through a sheath of ionized air, and propulsion system design and control (particularly for air-breathing systems)—all of which suggest areas for research and development to enable potential application by the Navy.

Fast-Kill Weapons

For both offensive (Sea Strike) and defensive (Sea Shield) missions, the ability of naval forces to deliver ordnance well within the maneuver or reaction

[11]Hypersonic missiles are explicitly called out as a Future Sea Strike technology in the Sea Power 21 vision.

time of the enemy is ultimately bound by basic kinematics (i.e., transit time and range). Since the velocities of conventional ballistic and rocket technologies are limited to small multiples of the speed of sound (i.e., < Mach 3) and the desired ranges of effectiveness continue to increase for deployment of weapons from safe zones, the situation is currently getting worse, not better, with time. The development of weapons with much greater velocity has the potential to reverse this trend.

Obviously, speed-of-light weapons employing lasers or microwaves suggest the potential for dramatic solutions. Low- and high-power lasers could revolutionize many aspects of offensive and defensive air-to-air and air-to-surface aerial warfare by providing precise, nearly instantaneous damage to targets. In addition, high-power microwave (HPM) weapons could disrupt all manner of electronics with little or no collateral damage. Additionally, HPM weaponry offers air-delivered, nonlethal control of humans on the ground as an adjunct to the traditional gunship weapons of today. Unfortunately, these technologies are also the most speculative, particularly within the size and weight constraints of aviation.

Very high velocity ballistic systems, such as electromagnetic rail guns, coil guns, and gas guns, are other examples of potentially applicable, less-than-the-speed-of-light technologies with advantages and technological risks different from those of hypersonic weapons, but with significant disruptive capability.

Heavy-Lift Air Transport[12]

The Marine Corps component of Sea Strike, STOM, depends critically on an ability to move both personnel and equipment on and off a sea-based platform to shore positions many hundreds of nautical miles inland with the same speed and volume as can be achieved using land transportation. Clearly this capability assumes a form of sea-based logistics far beyond the capacity of today's naval forces.

Sea Basing also envisions many situations requiring ship-to-ship or ship-to-shore movement of heavy equipment. Air lift, with STOVL and hovering capabilities, offers a potential solution. In addition, to support STOM the Marines require logistics support capable of high-speed horizontal flight, commensurate with the speed of the V-22 tilt rotor, which transports troops.

At least three general approaches are being considered to achieve these capabilities. While all are radical, the most conventional approach involves upgrading to the CH-53E, a large rotorcraft capable of transporting 16 tons of payload. Generically called the Heavy-Lift Replacement, it could see initial operational capability by 2015. A second approach, called the Air Maneuver

[12]For a longer discussion of this topic, see National Research Council, 2005, *Sea Basing: Ensuring Joint Force Access from the Sea*, The National Academies Press, Washington, D.C.

Transport, builds on the existing V-22 tilt-rotor technology by increasing the number of rotors; it would be capable of lifting as much as 20 tons but would require more development and would not be available until 2025. Finally, consideration is being given to an advanced class of very large, buoyant, lifting-body dirigibles. A version of this design is currently being evaluated by NAVAIR.

In the future, air-delivered logistics materiel may well be dropped at the precise point on the surface where it is needed, without any vehicle landing. Such an operation could be unmanned and autonomous, and hence carried out by vertical-lift UAVs. With such a capability, all facilities, infrastructure, vehicles, and so forth that land, off load, break down, reload, transship, unload, break down further, and distribute would no longer be needed. For example, food would be air delivered directly to the mess halls, ammunition to the weapons operator, and medicine to the field hospitals and aid stations.

Theoretical studies have been done of extreme heavy-lift vehicles. Able to lift 500 tons and travel at 100 knots, such a vehicle could move 1 million ton-miles per day into remote and unimproved areas. An entire regiment could be embarked on a small number of such vehicles with all of its equipment (including artillery, support, armor, and aviation), moving from its home base (no port or airfield required) to a remote site without forward logistics. It could land on unimproved forward areas, beach areas, or water.

Intelligent Combat Information Management

Naval aircraft need to be able to manage and display combat information in real time—that is, to prioritize and systemize the volumes of information generated both on board the aircraft and by other participants all drawing from and contributing to a single agile network. This capability would provide the pilot with instant access to all aircraft navigation, communications, sensor, display, self-defense, and weapons systems as well as automate many of the functions and lower-level decisions that the pilot makes today to enhance situational awareness and avoid information overload.

Ultimately, a combat information management system of the future could be envisioned, for the most part, to be "autonomic," that is, to run in the background, with little or no overt crew intervention needed. Autonomic information systems would organize and control the sensors, get the data, and build the picture automatically, anticipating the needs of the crew. They would pull the relevant military objects from the picture and convert them into prioritized and synchronized target queues. They would pair the sensors and weapons to the targets and construct the execution order. They would assess the results and update the picture. To achieve these autonomic systems would require bandwidth, algorithms, and processing power. Many of the supporting technologies for autonomic information systems exist in the laboratory today. Implementation of autonomic systems would get naval aviation closer to the goal of having pilots be advised in advance and human control exercised by exception or consent. This capability would minimize the chance that a target or necessary support function would ever become "critical" or "sensitive."

Omniscient Intelligence

The most vexing problem today is identifying the next target (what it is and where), which, if successfully attacked, would produce the largest desired effect. At some point in the future, it is expected that the persistent availability of sensors would make possible knowing with confidence what the next targets should be and where they are. Without question, achieving this level of "omniscience" would completely change the way naval forces operate.

The ability to sense virtually everything of interest in an environment, reduce the resulting raw data to critical information, and communicate that information to an autonomous system or human/machine system that would ultimately take action would provide a tremendous increase in capability. Examples include vastly improved reliability of systems as a result of condition monitoring and maintenance as needed (Sea Basing); awareness of all significant objects, friend and foe, in an area of interest (Sea Shield and Sea Strike); and grid processing for mission planning/conduct (FORCEnet)—all based on persistent/ad hoc networks that are self-forming, self-aware, and self-healing.

Tremendous strides have been made in recent years in computing, sensors, communications systems, and software to enable these capabilities. Further developments in these areas would lead to revolutionary improvements supporting all the pillars of Sea Power 21.

FINDINGS

- The concepts expressed in Naval Power 21 reflect a farsighted, aggressive, and challenging vision of future naval warfare for which neither a strategic operational plan nor a detailed implementation plan yet exists. Thus, capability needs and gaps for naval aviation have not yet been formally identified.
- Sea Power 21 and Naval Power 21 are revolutionary concepts envisioning how the Navy and the Marine Corps will fight future wars using a network-centric operational construct and architectural framework called FORCEnet.[13] FORCEnet is still in the early stages of development, and a complete operational and architectural design does not yet exist. Since a strategic plan that identifies concepts of operation and capabilities that leverage FORCEnet to gain superiority over potential adversaries had not been articulated by the Navy at the time of this

[13]For a longer discussion on this topic, see National Research Council, 2005, *FORCEnet Implementation Strategy*, The National Academies Press, Washington, D.C.

study, the committee used only the concepts embraced in Naval Power 21 to guide its determination of needed capabilities and gaps for naval aviation in the future relating to FORCEnet.

- NAVAIR currently lacks a naval aviation strategic plan that identifies capability gaps and technology development needs. A technology development plan established in cooperation with ONR does not exist. NAVAIR and ONR acknowledged the lack of a strategic plan for naval aviation's role in Naval Power 21. Both agreed that a naval aviation strategic S&T plan was essential, and both agreed to remedy the situation. As this study was being finalized, NAVAIR drafted the *Naval Aviation Vision 2020* document.[14] The committee believes this is a step in the right direction in forming the basis for such a strategic plan.
- NAVAIR's current strategy of exploiting near-term gains is manifested at ONR by an almost vanishing level of funding for basic research (level 6.1) allocated to naval aviation topics. The relationship between NAVAIR and ONR does not appear to be strategic, and the two organizations do not seem to form a collaborative team, despite the expenditure of substantial S&T funds (levels 6.2 and 6.3) on naval aviation topics at ONR.
- Current ONR planning appears to be largely ad hoc, with unclear goals against which to assess progress or ultimate value. The committee was unable to assess the relevance of current naval aviation S&T programs funded by ONR or their completeness in furnishing needed capabilities for Naval Power 21. No institutional process is currently in place at ONR to create or contribute to a vision of naval aviation for the future.
- The perception of disarray in ONR aviation-related S&T was exacerbated by the large fraction of naval aviation programs that are funded by congressional largess and have little connection to Navy objectives.
- The strategic S&T planning processes of both the Army and the Air Force contain much that the committee believes could help the Navy in its planning process.

RECOMMENDATIONS

- To enable the capabilities for naval aviation operations as envisioned in Naval Power 21, the Chief of Naval Research, in partnership with NAVAIR, should lead the development of a naval aviation strategic S&T plan.

---

[14] As a result of a cooperative effort sparked by the present study, NAVAIR and ONR have issued the document *Naval Aviation Vision 2020* (see VADM James M. Zoortman, USN, Commander, Naval Air Forces; VADM Walter B. Massenburg, USN, Commander, Naval Air Systems Command; and RDML Thomas J. Kilcline, Jr., USN, Director, Air Warfare Division, 2005, *Naval Aviation Vision 2020*, Naval Aviation Enterprise, Department of the Navy, Washington, D.C. Available online at <http://www.nae.cnaf.navy.mil/demo/main.asp?ItemID=12>. Last accessed on September 30, 2005).

As this study was being finalized, the Commander of Naval Air Forces, the Commander of NAVAIR, and the Director of the Air Warfare Division in OPNAV created the document *Naval Aviation Vision 2020*, which can provide a basis for the development of this strategic S&T plan. This plan should be updated annually in synchronization with the Planning, Programming, Budgeting, and Execution System process. It should also be considered in the naval aviation S&T plans of the Army and the Air Force.

- ONR should establish a formal process for the identification of key S&T approaches that will identify and address naval aviation capability gaps. A methodology should be developed for analyzing options and selecting preferred approaches based on a systems perspective that includes technology trade-offs, maturity, risks, cost, impact, and so on. A methodology should also be developed for connecting novel concepts and potential breakthroughs into a naval aviation strategic S&T plan such that they receive attention, undergo development, and have a path into the acquisition domain.
- ONR should consider the S&T planning processes used by the Army and the Air Force as a source of potential guidance in developing a naval aviation strategic S&T plan.
- The naval aviation strategic S&T plan developed by ONR and NAVAIR should identify capability needs and current capability gaps that technology development will address. Technological progress and investment levels should be tracked each year for every major and minor technology category.
- As ONR develops a naval aviation strategic S&T plan, consideration should be given to the following disruptive aviation capabilities, each of which can be traced to at least one of the four components—Sea Shield, Sea Strike, Sea Basing, and FORCEnet—of Naval Power 21:

—Multispectral defense,
—Unmannned air operations,
—Hypersonic weapons delivery,
—Fast-kill weapons,
—Heavy-lift air transport,
—Intelligent combat information management, and
—Omniscient intelligence.

Science and technologies in which ONR could pursue advances to enable each of these capabilities are discussed in Chapter 3.

3

Science and Technology for the Disruptive Capabilities

Just as the list of disruptive capabilities discussed in Chapter 2 is by no means exhaustive of the possibilities that ONR could pursue, so too the S&T opportunities identified in this chapter are not exhaustive. Committee members used their experience and expertise to provide a high-level assessment and to suggest where emphasis should be placed with respect to investments in Discovery and Invention (D&I) programs (6.1 and early 6.2) and Exploitation and Development (E&D) programs (late 6.2 and 6.3). The D&I programs tend to be longer-term, higher-risk efforts that, for the purposes of this study, tend to fall into the 2011 to 2025 time frame. The E&D programs tend to be shorter term with reduced technical risk, and are aimed for early insertion and transition to the fleet in the 2007 to 2010 time frame.

The recommended S&T opportunities, wherever possible, are categorized as (1) naval unique (required only by naval missions), (2) naval essential (important for naval and non-naval missions), and (3) naval relevant (useful for naval and non-naval missions).

MAPPING TO FUNCTIONAL SCIENCE AND TECHNOLOGY AREAS

The committee identified seven S&T functional areas as highly relevant to naval aviation: avionics technology, sensors, propulsion and power, structures and materials, human systems integration, survivability, and the core technologies of aerodynamics and modeling and simulation. For each disruptive capability the committee identified the S&T areas it believes require a focused development effort in order to make the capability a reality in the future. Table 3.1 shows

TABLE 3.1 Mapping of Disruptive Capabilities to Functional Science and Technology Areas

Capabilities	S&T Area by Function[a]						
	Avionics	Sensors	Propulsion and Power	Structures and Materials	Human Systems Integration	Survivability	Core Technologies[b]
Multispectral Defense							
Active stealth	√ √	√		√		N/A	√
Active countermeasures	√ √			√		N/A	
Unmanned Air Operations							
High-altitude, long-endurance (commmunications, reconnaissance)	√ √		√ √	√		√	
Small-/intermediate-sized UAVs			√ √				
Tactical, high-performance (e.g., UCAV)	√ √		√		√	√	
Micro-UAVs (man-portable)	√	√	√ √	√	√		√
Hypersonic Weapons Delivery							
Air breathing		√	√ √	√ √		√	√
Rocket		√		√ √			
Fast-Kill Weapons							
Soft-kill lasers							√
Structural-kill lasers (e.g., air-breathing laser)			√				√
High-power microwave (very soft kill)			√				√

Heavy-Lift Air Transport							
Next-generation rotorcraft/VSTOL			√	√√	√	√	√
Wing-in-ground effect	√						
Dirigible				√√			
Intelligent Combat Information Management							
Autonomic (track automation)	√√				√√		
Autonomy	√√						
Omniscient Intelligence							
Persistent/ad hoc network	√						
Grid processing for mission planning/conduct	√				√		√
Reconfigurable sensor nodes	√						
Robust sensing		√					

NOTE: UAV, unmanned aerial vehicle; UCAV, unmanned combat air vehicle; VSTOL, vertical short takeoff and landing.

[a]A "√" symbol indicates the committee's view that a strong, well-planned, focused technology effort, requiring an order-of-magnitude improvement or breakthrough in fundamental or implementation technology, is required. A "√√" symbol indicates that an even more significant investment and/or improved S&T program is required to realize the corresponding desired capability. The absence of a symbol does not necessarily imply that the committee thinks no effort is required, but rather that the priority is lower relative to other areas.

[b]Aerodynamics and modeling and simulation.

the mapping of each capability and its subsets to the functional S&T areas. Discussion regarding the mapping rationale and suggestions for particular S&T areas of focus within each functional technology area are presented below.

Avionics Technology[1]

The superiority of U.S. naval aviation and the realization of the "know quickly and act decisively" concept are critically dependent on situational awareness, excellent communication and coordination, rapid precision strike, and greater employment of unmanned aircraft. Advancement of avionics technology is integral to each of these concepts. The committee believes that significant technology challenges exist and that ONR should play a leadership role in the development, experimentation, and evaluation activities conducted to mature these new capabilities.

At this time, however, ONR appears to be disengaged as a leader and sponsor of avionics technology development. In a 2001 study the Naval Studies Board's Committee for the 2001 Assessment of the Office of Naval Research's Aircraft Technology Program observed that "ONR is exiting the field of avionics technology development" and asserted that, with minor exception, the avionics work at ONR was focused on engineering "fixes" to current systems rather than investing in technology for the next generation.[2] The current committee has not seen evidence that this trend has been reversed.

Although the 2001 report was more favorable in its assessment of ONR's work in autonomy for UAVs, this committee notes that the Navy's previous centerpiece program, UCAV-N, has been moved to the Defense Advanced Research Projects Agency (DARPA) as part of the Joint Unmanned Combat Air System program. In addition, the committee was informed that the investment in autonomous technology at the 6.1 D&I level is to be reduced from $4 million to $2.5 million for fiscal year 2005.[3]

The committee recognizes that great advances in computing speed, memory density, wireless networks, and distributed computing are being driven by commercial market forces; thus, although these are critical enablers of future naval avionics, they are likely not appropriate investment areas for ONR. Further, significant work on aircraft avionics relevant to several of the identified disrup-

---

[1]For a lengthier discussion on this topic, see National Research Council, 2004, *Evaluation of the National Aerospace Initiative*, The National Academies Press, Washington, D.C.

[2]Naval Studies Board, National Research Council. 2001. *Assessment of the Office of Naval Research's Aircraft Technology Program*, National Academy Press, Washington, D.C.

[3]Clifford W. Anderson, Office of Naval Research, "Strike Technology Division (ONR 351) Basic Research Program," presentation to the committee on September 28, 2004.

tive capabilities is being executed in the other military Services.[4] Thus, a systematic review of avionics technology to identify naval-aviation-unique gaps is needed. ONR should be a leader and principal focal point for this systematic evaluation within the Navy. ONR should then be the principal sponsor of a cohesive program for the development of new technologies to meet the identified gaps.

Although ONR is in partnership with the larger aviation community, the committee believes that the advanced avionics technologies involved in three of the identified disruptive capabilities—namely, multispectral defense, unmanned air vehicles, and intelligent combat information management—will be central to the future of naval aviation and therefore of high interest to ONR. Specific comments on these areas follow.

Multispectral Defense

Current low-observable (LO) technologies rely primarily on shaping and materials. These technologies are well understood, and their use has proliferated in many countries. Naval aviation can expect to encounter today and in the future adversarial air platforms with varying degrees of LO signature. The challenge for naval aviation will be to exploit all platform signatures—first for its own survivability and second to counter any adversary. Shaping and materials can achieve significant reductions in signatures, but they will not be sufficient for naval aviation in the future. Current technology permits tactically significant reductions in both radio frequency (RF) and infrared (IR) signatures.

Visual and acoustic signatures are very important to low flyers, and "all-aspect" stealth is very important to high flyers. Both need dedicated S&T efforts.

The next LO frontier is visual signature reduction. It could possibly involve active systems that change color or hue, reflectivity, or emittance. Conceptually, a visual LO system blends vehicle visual signature information with that of the background. If the vehicle is moving, this visual LO system must be very agile, with extremely fast response rates. An air platform with an active visual LO signature reduction system could be a really disruptive capability.

Active systems would have to be employed by naval aviation in order to reach the signature levels necessary to win against future adversaries. Active systems would require advanced high-speed electronics with robust sensors and detectors, high-speed networks, and advanced processing algorithms.

---

[4]For example, the Air Force has efforts in lasers for fast kill, and the DOD, with Army lead, is working toward developing and testing heavy-lift air transport vehicles.

Unmanned Air Operations[5]

Significant advancements in the level of autonomy will be required to improve the effectiveness of UAVs, which will be required to function collaboratively in a complex and threat-filled environment and in conjunction with manned aircraft. Operating such vehicles from a rolling ship deck as part of a mixed wing of manned and unmanned systems is unique to naval aviation.

To deal with these challenges, these UAVs must process sensor data, identify unexpected threats and opportunities, perform dynamic planning, communicate, and execute their missions with limited human intervention. Because of these difficult requirements, the committee recommends a reinvigorated ONR focus on high-level intelligent autonomous technology[6] that enables the following:

- Constellations of self-organizing and self-directing UAVs with human "control by exception" for automatic surveillance, reconnaissance, targeting, and attack;
- Development of an integrated digital system for providing automated communications (both RF and optical) and control between swarms of UAVs and ground stations to include networking of imagery, signals intelligence, Global Positioning System geolocation through sequencing antennas, and software-controlled radios;
- Self-contained automatic carrier landing capability with very-low-probability-of-intercept emissions from the aircraft platform (either manned or unmanned) and an aircraft carrier that does not have to emit a signal; and
- Reduced need for human involvement in UAV operations.

In addition to the need for a greater emphasis on technologies that enable autonomous operations, a principal challenge and a key element of the successful operation of multiple UAVs in a complex battle space will be the achievement of dramatically enhanced situational awareness for the remote "pilot-commander." Such situational awareness will be crucial during strike operations (especially decisions about weapons release) and the naval-unique close-quarters operations of manned and unmanned systems on the carrier deck. Consequently, ONR should pursue a technology development program that enables:

- Simultaneous operations on carrier decks with manned and unmanned aircraft characterized by robust situational awareness, fault tolerance in both systems, and a significantly improved command and control architecture; and

---

[5]For more discussion on this topic, see National Research Council, 2005, *Autonomous Vehicles in Support of Naval Operations*, The National Academies Press, Washington, D.C.

[6]National Research Council. 2005. *Autonomous Vehicles in Support of Naval Operations*, The National Academies Press, Washington, D.C.

- "In-cockpit-like" situational awareness for the remote UAV "pilot-commander" through the robust low-latency downlink of critical sensor and video data and the development of advanced synthesis and presentation technologies such as virtual reality techniques for visual, aural, and tactile feedback.

Operating UAVs in conjunction with manned aircraft on an aircraft carrier deck is perhaps the most naval-unique challenge facing naval aviation. Such operations will significantly disrupt carrier deck procedures as they are known today. Mixed-wing operations on the carrier deck will require a significant infusion of technology to provide robust situational awareness to all participants, fault tolerance in both the individual systems, and the command and control architecture needed to achieve efficient and safe operations.

Positive control of the unmanned aircraft will be required by the carrier deck personnel who currently communicate with pilots via hand signals. Methods are needed for the carrier deck personnel to instruct UAV aircraft, confirm receipt and understanding of these instructions, and interrupt operations. The technology could include a vision system for the UAV, a data terminal for personnel, or remote "pilot" operator stations with advanced situational awareness capabilities.

Intelligent Combat Information Management

Naval aircraft need a new digital high-speed intelligent combat information management and display system (IMDS) that prioritizes and synthesizes the volumes of information generated on board the aircraft and provided by other participants on the Navy FORCEnet.[7] This system would be the principal pilot interface to all aircraft navigation, communications, sensor, display, self-defense, and weapons systems. It would automate many of the functions and lower-level decisions made today by the pilot to enhance situational awareness and avoid information overload.

An IMDS would be enabled by a fully integrated avionics architecture (a federated system of sensors, displays, processors, and countermeasure subsystems) and would require significant supervisory intelligence as it automates/manages aircraft functions. The on-board IMDS would exploit an underlying Internet Protocol (IP) connection to FORCEnet and automate the pushing and pulling of mission-critical data from the network to meet higher-level mission objectives. It would interface to other nodes: aircraft, UAVs, command and control, databases, sensors, weapons, and so on.

---

[7]ADM Vern Clark, USN, Chief of Naval Operations, and Gen Michael W. Hagee, USMC, Commandant of the Marine Corps. 2005. "FORCEnet: A Functional Concept for the 21st Century," Department of the Navy, Washington, D.C., February.

Research and development into IP-based, high-bandwidth, optical aircraft intranet structures for on-board data management systems could greatly reduce the cost and weight of the avionics infrastructure. Additionally, cost-effective ways are needed to retrofit current analog/1553 digital bus aircraft with state-of-the-art digital high-speed architectures.

The IMDS would also be an advanced operator-vehicle interface to provide enhanced situational awareness and higher-level interaction with the pilot. To create this system and its component technologies requires human factors research into information assimilation and cognitive decision tools. This subject is covered in some detail in the section below titled "Human Systems Integration."

Sensors

Achieving Sea Power 21 and Marine Corps Strategy 21 is critically dependent on the pervasive use of sensors and their capabilities for intelligence, surveillance, and reconnaissance; for threat warning; for targeting; for weapons' guidance; and even for such mundane things as controlling the physical configuration of aircraft in flight and detection of failing or failed components for condition-based maintenance or for modes of fault tolerance and self-healing. The range of available sensors is extremely large and diverse—encompassing sophisticated electromagnetic modalities from ultralow microwave frequencies through the conventional microwave bands to millimeter wave (MMW) regions and on into the optical ranges of the infrared, the visible spectrum, and beyond, as well as inertial sensing, the detection of chemical and biological agents, acoustic and seismic sensing, and environmental sensing of temperature, humidity, wind conditions, and so on.

Table 3.2 represents a partial listing of Navy-relevant sensor technologies and suggests the extremely wide range of topics that the term "sensors" encompasses.

The key to the success of network-centric war-fighting concepts in naval aviation is the cooperation of multiple sensors and sensor platforms and the successful implementation of multisensor fusion, exploiting the information from multiple sensors distributed throughout the battle space to create, in real time, continuous and complete battle space awareness. The Navy's Cooperative Engagement Concept has shown the effectiveness of a such a distributed sensor (i.e., radar) configuration, although the specific implementation, which places enormous loads on the communications (e.g., directional point-to-point, high power, very wide bandwidth), is probably not a model for future, more general, FORCEnet implementations.

From the point of view of the disruptive capabilities listed in Table 3.1, it is clear that sensors, in one form or another, must play a crucial role in any new concept that arises. However, although this is undoubtedly true, Table 3.1 does not call out sensors for enhanced attention either because the committee thought

TABLE 3.2 Individual Sensor Technologies

Sensor Technology	Topics Encompassed
Radar	Conventional wavelengths: SAR, MTI, and so on; all-weather MMW, active and passive for visual-quality imaging through obscurants; very-low-frequency SAR for foliage and ground penetration
Optics	Passive Infrared, visible, multi-/hyperspectral, and so on Active Imaging LADAR Foliage and water penetration—LADAR, blue, tunable Countermeasures—tunable infrared laser Configuration issues Hypersonic aerodynamics constraints
Chemical weapons of mass destruction	Local Long range
Biological weapons of mass destruction	Local Long range
Inertial	MEMS for low cost, small size
Acoustic, thermal pressure, humidity, and others	Air vehicle health/maintenance sensors for condition-based maintenance, persistence via fault tolerance and self-healing, environmental sensing, and so on

NOTE: SAR, synthetic aperture radar; MTI, moving-target indicator; MMW, millimeter wave; LADAR, laser detection and ranging; MEMS, microelectromechanical systems.

that the technology is being adequately developed as a whole by government, contractors, and universities or because the ONR programs on sensors currently appear to be on a reasonable path to needed future capabilities. New sensor technology developments do not seem to be required for the particular disruptive capabilities identified in the previous chapter.

Indeed, in general, ONR's radar and optical sensor work seems mainstream—that is, comparable to the state-of-the-art work available in the government, university, and contractor communities as a whole. In particular, ONR's (Code 31) work on multifunctional RF antennas (i.e., common hardware for radar and communications at multiple frequencies) is novel and of great importance for future platforms that implement FORCEnet interconnections—for the resolution of cross-platform radar/communication interference issues and for just plain real estate reasons.

Of course, there are always sensor challenges to be met and new ideas and opportunities to exploit in every application. Below are some comments on the technical challenges inherent in the disruptive capabilities listed in Table 3.1 as requiring focused efforts in sensors.

Sensors for Multispectral Defense

For multispectral defense, the challenge is to characterize the background along the line of sight of the threat so that a UAV can respond by actively altering its signature to minimize its contrast with the background. Although this capability is not difficult in static situations, the dynamics of flight (speed, maneuvering, angles, etc.) will stress sensor response capabilities dramatically.

Sensors for Micro-UAVs (Man-Portable)

The very small sizes and severely limited on-board power capacity of micro-UAVs create special sensor challenges that, although they have already been addressed for modest-sized UAVs, require a dedicated S&T effort.

Sensors for Hypersonic Weapons Delivery

Sensor aperture compatibility with aerodynamic requirements is a very difficult problem as both may be compromised. Forward-pointing sensor apertures are not easily made compatible with the slender pointed shapes and shock-wave-withstanding structures dictated by hypersonic aerodynamics. Even conventional radar-guided missiles (e.g., Hawk, Patriot) suffer difficult bore sight errors because they must receive microwave signals through aerodynamically shaped radomes. Full imaging with optical sensors, which is desirable for high-resolution, precise, end-game tracking of a hypersonic missile, could be seriously compromised by the need to look through a complex pointed optical element (i.e., the tip of the hypersonic vehicle) or through a structure whose shock-wave resistance might vary considerably with angle of attack. Side-looking apertures could more easily be made compatible with the aerodynamics, but without a direct forward field of view a vehicle or a missile must resort to complex flight paths to compensate for the unfortunate blind spot that results.

Sensors for Omniscient Intelligence

The ability to reliably "see through" weather, obscurations, and foliage and to penetrate water is desired for many future naval applications. While optical systems are limited to clear-air environments, conventional radars and MMW systems are relatively insensitive to weather, smoke, fog, and so forth, and MMW imaging is of particular interest for both active and passive short-range, high-resolution imaging in obscured environments.

Laser detection and ranging (LADAR), with its high-resolution three-dimensional geometric imaging capabilities for looking through holes in the foliage canopy, and ultralow-frequency microwave synthetic aperture radar, which is barely affected by dry leaves, show promise for foliage penetration. Both are currently under active investigation by the Army Research Laboratory and DARPA.

"Blue" LADAR technology can penetrate to various depths in the ocean typically measured in a few "extinction depths." Although this capability is advantageous in some parts of the ocean, it is inadequate in many littoral environments, harbors in particular.

Other Sensor Technology Issues

While the disruptive capabilities discussed above do not seem to require the development of completely new sensors beyond current state-of-the-art capabilities, there are a number of sensor technology issues that Navy S&T should be monitoring or actively involved in. Three of the most interesting are discussed briefly below.

Digital Creep. Most sensor systems generate their information via an analog measurement of the environment and then conversion of signal information into a digital format for processing and subsequent extraction of information. As the performance of digital chip technology continues to increase exponentially, the trend in sensor systems is to convert the signal to digital closer and closer to the front end of the system and to exploit the compact storage and computational power inherent in modern digital circuits to provide on-board signal processing and information extraction—thus producing very capable "smart" sensors.

Military optical systems have already become largely digital, with sophisticated, high-density focal plane arrays, on-board and sometimes on-chip signal processing, and the like. Radar, too, has been heading in the same direction (with DARPA encouragement), albeit more slowly because the analog-to-digital (A/D) challenges of directly digitizing microwave signals are enormous, given the requirements for very high speed, power-hungry sampling. Yet the benefits of practical (and affordable) A/D capability—including freedom from environmentally induced drift, the possibility of microwave-insensitive wide-band digital phase shifting, and the physical benefits of replacing microwave wave guides by fiber-optic communication links—could be significant. Navy S&T should closely follow the trend toward digitalization and consciously apply it to other than optical and radar sensor modalities.

Nanotechnology. Nanotechnology has much to offer for enhanced sensor capabilities. Operating at nano-level (10^{-9}) dimensions, it is becoming possible to construct many artificial materials that do not appear in nature and that could have novel and unusual mechanical, electrical, or optical properties. Recently, oriented carbon nanotubes, for example, which can be made in insulating or semiconducting forms, have been shown to have the capability to act as optical detectors. Single-electron transistors, constructed via rapidly developing nanotechnology techniques, also offer promise for future novel applications to sensors.

Perfect Imaging. Thirty or more years ago it was noted that if a material with a refraction index equal to –1 could be found, then perfect imaging, free of diffraction-induced wavelength limitations, could be implemented.[8] Materials with a dielectric constant ε and a magnetic permeability μ both equal to –1 have not been found in nature, and so the concept of making "superlenses" has not yet been demonstrated. A few years ago, however, artificial materials consisting of oriented wires (for ε control) and embedded rings (for μ control) were constructed[9] with a resulting refractive index of –1 at microwave frequencies, and the possibility of "perfect imaging" or superlenses has been demonstrated. An active research community has sprung up to exploit this idea.[10]

With the potential to bypass the diffraction limitations of radar antennas, so-called perfect imaging could offer major advances in radar sensors, particularly for MMW imaging from small platforms where physical limitations restrict usable aperture size. This new approach is a high-risk effort but is worth watching closely.

Propulsion and Power

Currently, the ONR propulsion program is rather limited and evolutionary in character. To support the Navy in Sea Power 21 and Marine Corps Strategy 21, ONR needs to stretch its vision farther into the future and start work on some revolutionary propulsion technologies. The committee suggests that a good place to start would be the technologies needed to enable the identified disruptive capabilities areas of UAVs, hypersonic weapons delivery, and heavy-lift vehicles.

ONR participates with the Air Force, General Electric, and Pratt & Whitney

[8]V.G. Veselago. 1968. "The Electrodynamics of Substances with Simultaneously Negative Electrical and Magnetic Permeabilities," *Soviet Physics USPEKHI*, Vol. 10, pp. 509-514.

[9]R. Shelby, D.R. Smith, and S. Schultz. 2001. "Experimental Verification of a Negative Index of Refraction," *Science*, Vol. 292, p. 77.

[10]See, for example, the work of J.B. Pendry at the Blackett Laboratory, Imperial College, London, England, and of David R. Smith at the University of California, San Diego.

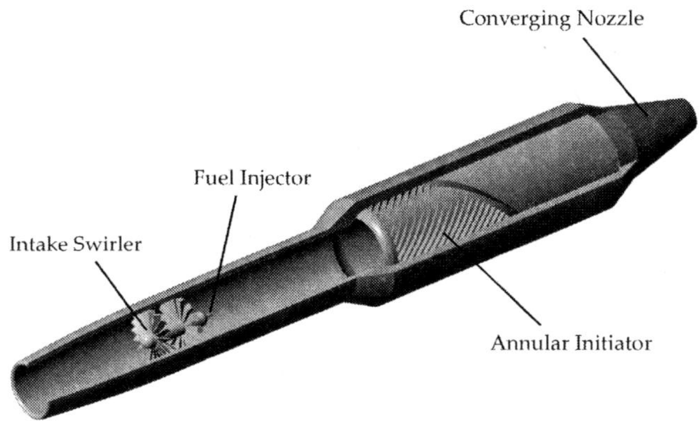

FIGURE 3.1 Pulse detonation engine. SOURCE: Office of Naval Research.

in the Integrated High-Performance Turbine Engine Technology (IHPTET) and the Versatile, Affordable, Advanced Turbine Engine (VAATE) programs. These programs integrate various materials and component technologies and demonstrate technologies in a realistic engine environment that is well instrumented. The IHPTET program is intended to achieve a 20 percent improvement in fuel consumption and 30 to 35 percent improvements in thrust, range, payload, and maintenance cost. These improvements would be substantial evolutionary gains accomplished over a mid-term period. The IHPTET program does not involve high-risk, potentially disruptive technologies but rather is a more conservative attempt at advancement of capabilities.

Jet engine noise reduction is another area of interest to ONR. The specific plans for this area have not been made clear. A program that addresses exhaust noise as well as compressor noise should be pursued.

One ONR research program addresses the important problem of jet engine flame-out; another examines the use of active control of combustion processes for enhanced mixing and performance. Active combustion control is a promising research area from which several valuable enabling technologies may develop. Research in this area of active control should continue and should be broadened and deepened as a core activity in combustion.

ONR does have a research effort in the area of pulse detonation engines.[11] See Figure 3.1. This relatively simple and inexpensive engine design could sub-

---

[11]G.D. Roy, S.M. Frolov, A.A. Borisov, and D.W. Netzer. 2004. "Pulse Detonation Propulsion: Challenges, Current Status, and Future Perspective," *Progress in Energy and Combustion Science*, Vol. 30, No. 6, pp. 545-672.

stantially reduce costs for weapons delivery systems (e.g., cruise missiles that currently use expensive gas turbine jet engines). This is a high-risk controversial concept for several reasons. It would have an inefficient thermal cycle since entropy-producing discontinuous compression waves are used to compress the combustible mixture. Also, the required mechanisms to cause detonation of fuel-spray/air mixtures in tubes no longer than a few meters can lead to complexities, volumes, weights, and costs that were not originally envisioned. For example, long nozzles, electric arcs, and rocket propellants have been considered.

The search for less costly weapons delivery should be broadened beyond the pulse detonation engine. One approach would involve studies of potential schemes that use rockets or ramjets (including hypersonic ramjets) for sequential ballistic/glide or powered flight trajectories that enable a missile to survive defenses as it flies to its target. Studies should address both the cost and the feasibility of this approach. A jet-assisted takeoff-boosted ramjet should be considered as a low-cost propulsion system. Two-stage vehicles could be studied in which the first stage has a turbojet engine and the final stage has a rocket or ramjet engine. Of course, the intention is to recover the first stage. Recovery mechanisms could be studied for cost and feasibility. The first stage might be a UAV that flies back to the carrier; it could be a UAV seaplane, or it could fall with a parachute into the sea. ONR should engage in fundamental research related to novel concepts that offer the possibilities of such cost reductions.

Hypersonic flight with continuous combustion of liquid fuels is an area that requires further study for optimization of the supersonic combustion process. Although the Air Force has a program on hydrocarbon-fueled scramjets,[12] relatively little is known about the scientific foundations for injection, atomization, and vaporization of liquids into a supersonic air stream. Emerging concepts that bypass the problems of supersonic combustion could also be examined (e.g., crossed-field electromagnetic deceleration of incoming air and acceleration of product gases).[13]

There is a need for a heavy-fuel engine for smaller rotary-wing UAVs. Currently, only light-fuel engines are available in certain size ranges. Efforts should not be restricted to scale-down of existing propulsion technology. New concepts should be considered.

ONR is currently not engaged in the aeronautics of long-range, long-endurance, small (10 cm to 1 m) fixed- or rotary-wing UAVs. These would require highly efficient engines (or other power sources) that overcome the adversities of

---

[12]R.S. Fry. 2004. "A Century of Ramjet Propulsion Technology Evolution," *Journal of Propulsion and Power*, Vol. 20, No. 1, pp. 27-58.

[13]D.W. Riggins. 2004. "Analysis of the Magnetohydrodynamic Energy Bypass Engine for High-speed Airbreathing Propulsion," *Journal of Propulsion and Power*, Vol. 20, No. 5, pp. 779-792.

increased surface-volume ratios at small scales (e.g., increased heat loss per unit of energy converted, increased friction dissipation per unit of energy converted). Mere scale-down of existing designs will not work; new concepts and the associated research and enabling technology development are needed. A major new effort in miniature propulsion systems should be established.

There is also the need to study aerodynamics issues for miniature UAVs. In some new concepts, propulsion and aerodynamics might become more strongly integrated. See the "Core Technologies" section of this chapter for further discussion.

The naval interest in airborne laser weapons may have synergy with propulsion and airborne power needs. Lasers are very energy inefficient in that only a small fraction of the energy from the laser power source appears in the photon beam. Basic research to explore uses of the remaining energy for the purpose of propulsion or other airborne power needs should be undertaken.

ONR has considered propulsion and power as critical technology areas. Nevertheless, there is a strong need to strengthen and broaden S&T efforts in this domain.

Structures and Materials[14]

Navy aircraft do everything that land-based aircraft do—but in a more hostile environment and under more adverse conditions, demanding readiness to respond in all sea states, in all weather conditions, and from both large and small ships. Catapult takeoff and arrested landings impose high-impact structural loads that threaten either fracture or low-cycle fatigue failure of landing gear and other structural components. The marine environment exposes all structural components to extremes threatening both corrosion and stress-assisted failures. Limited storage space leads to design options that differ markedly from those of comparable land-based aircraft, and these design differences impose materials selection options that may differ radically from those for their land-based counterparts.

These differences in environment and basing have led to selections of materials and design that make naval aircraft unique, and this pattern does not seem likely to change for UAVs. Persistence, rapid turnaround, fuel efficiency, and similar systems requirements will continue to demand advances in the development of lightweight stealthy materials and structural design principles for these unusual materials. The specific applications in naval aviation systems may be unique, but the underlying principles and D&I programs will strongly overlap and should complement those engaged in by the other Service S&T programs.

---

[14]For more discussion on this topic, see National Research Council, 2003, *Materials Research to Meet 21st Century Defense Needs*, The National Academies Press, Washington, D.C.

These issues concerning materials and structures have been appreciated by ONR in the development of its S&T program in support of propulsion and power for next-generation naval aviation. The committee was briefed on that program and its strong participation in the interagency IHPTET and VAATE programs. By contrast, based on the briefings received by the committee, there does not appear to be comparable attention given to the materials and structural issues associated with the other elements of naval air vehicle platforms.

The committee recommends that ONR examine the technology areas listed in Table 3.3 as a way to enable and realize the identified disruptive capabilities. Table 3.3 also suggests improvement goals for both the 2007 to 2010 and 2011 to 2025 time frames. Each of the recommended technology areas is identified as unique, essential, or relevant to naval needs.

While structural research and development in other areas is still required (e.g., high-altitude, long-endurance UAVs), the committee thought that, given a limited budget, the greatest benefit would be achieved by focusing on the areas of hypersonic weapons delivery and heavy-lift air vehicles.

Advanced structures are seen as a key enabler in both hypersonic and heavy-lift air vehicles. Each disruptive capability area drives the required structural developments to two different extremes: lightweight/high-temperature capability (hypersonic applications) and lightweight/high-stiffness capability (all applications, especially for heavy-lift air vehicles). As with all naval aviation, affordability, producibility, and supportability in a maritime environment cannot be sacrificed in the name of higher performance. For these disruptive capabilities to be realized and introduced into fleet applications, a systems engineering process is required to ensure that all of the needed capabilities are balanced.

The following paragraphs address some options that should be included in such a process to lower the development risk. No exhaustive listing is attempted, and the specific topics selected should be identified after careful interagency discussions to clarify the desired breadth of an expanded program and the Navy-specific and Navy-relevant aspects of new investments. A more comprehensive NRC examination of materials for all future DOD needs[15] should be consulted as these programs are developed by ONR.

Structures and Materials for Hypersonic Weapons Delivery

In the area of hypersonics, key revolutionary developments (order-of-magnitude improvements) to achieve lighter weight, high-temperature capability, high strength, and producibility are needed to enable next-generation applications.

[15]National Research Council. 2003. *Materials Research to Meet 21st Century Defense Needs*, The National Academies Press, Washington, D.C.

Within this disruptive capability area, two distinct hypersonic applications have been identified: air-breathing-based and rocket-based air vehicles. Greater structural research and development (in conjunction with high-temperature materials) are required to enable the design and integration of the aircraft/missile inlets/exhausts, control surfaces, primary edges, and so on. These advanced structures must also be robust enough to enable optimum design flexibility for aerodynamics and aerothermal considerations for hypersonic flight.

In addition, attributes that the ultimate system may require, such as LO or subsurface launch, will further complicate the systems design and also underscore the need for early research and development in high-temperature, structurally embedded LO sensor windows and apertures and watertight high-temperature seals. Furthermore, low-cost hypersonic structures will be required for missile applications.

If the Navy leadership, through a detailed systems engineering process, determines that manned or unmanned hypersonic aircraft should be based on and return to ships, the hypersonic-capable airframe would have to incorporate low-speed handling for ship recovery. This could entail propulsion systems capable of not only hypersonic flight but also short takeoff and vertical landings.

Of all the issues needing attention before serious application of hypersonic speed in missiles or vehicles, high-temperature materials and manufacturing technology for their fabrication into parts remain at the top of the list demanding further D&I. This is not so much a naval-unique issue as it is a naval-essential one. While programs exist in other agencies, they are inadequate to address all issues at current levels of effort. Naval participation in such programs will not only accelerate the development of critical technologies, but will also bring the Navy to the table as part of the DOD "smart buyer" team when acquisition begins.

Ceramic Materials. Unlikely to have any immediate effect on vehicle structural members, advanced ceramics are already beginning to impact rocket and turbine engine exhaust systems and turbine blades, the former as solid elements, the latter as coatings. The high strength required at the high temperatures inherent in hypersonic missiles will undoubtedly be achieved using ceramic materials. Many of these materials either do not yet exist or will require further S&T before applications can be considered by designers.

Metals. High-temperature materials, including molybdenum-based alloys, attract continuing attention for engine applications, promising increased thermal efficiency if oxidation problems can be overcome with suitable coatings. Significant investment is being made in the development of these materials under the IHPTET and VAATE programs. However, recent advances in the application of computational tools to the development of improved structural steels have led materials

TABLE 3.3 Technology Opportunity Areas to Enable Disruptive Capabilities

Disruptive Capability Area	Technology Area Opportunities
Unmanned aerial vehicles (UAVs)	Efficient lifting surfaces/morphing structures
UAVs/hypersonic vehicles	Structurally embedded apertures
UAVs	Extremely lightweight structures
UAVs	MEMS-based active control of lifting surfaces
UAVs	Lightweight self-monitoring and self-healing structures
UAVs/hypersonic vehicles	Advanced low-observable sensor windows with high-temperature capability
Directed energy	Lightweight structures hardened against directed-energy weapons attack
Directed energy	High peak power with rapid-heat-dissipation-capable structures
Heavy-lift air vehicles	Very lightweight ballistic-projectile-resistant armor/structures
Heavy-lift air vehicles	Active reconfigurable rotors and control surfaces for optimized flight
Heavy-lift air vehicles	

NOTE: MEMS, microelectromechanical systems; VLO, landing gear operational speed.

Suggested Technology Improvement Goal (2007-2010)	Suggested Technology Improvement Goal (2011-2013)	Naval Relevance (Unique or Essential)
Greater lift/drag—long loiter (>24 hr)	Optimized loiter vs. flight profile—very long loiter capability (7 days)	Essential
Reduced structural integration weight with VLO compatibility for subsonic/supersonic flight applications	Reduced structural integration weight with VLO compatability for hypersonic flight applications	Essential
Lightweight structures applicable to small UAVs	Ultra-lightweight structures for nano-UAVs	Essential
Very lightweight, highly reliable actuators for small UAVs	Ultra-lightweight, highly reliable actuators for nano-UAVs	Essential
Health monitoring of primary structures with self-reporting of maintenance actions required	Self-healing structures	Essential
Affordable sensor windows compatible with sensor requirements for subsonic/ supersonic flight	Affordable sensor apertures compatible with sensor requirements for supersonic flight	Essential
Lightweight structures capable of withstanding directed-energy weapons attack		Essential
	Lightweight structure with the ability to withstand high peak power pulses while dissipating the heat generated without degradation of the primary structure's strength	Essential
Lightweight structures that help reduce overall weight, enabling 500- to 1,000-nm ranges with typical helicopter flight profiles	Lightweight structures that help reduce overall weight, enabling 1,500- to 2,000-nm ranges with faster flight profiles than existing helicopters (tilt-rotor-like)	Essential
Lightweight armor protection coatings/structure	Integrated armor protection provided by primary air vehicle skins	Essential
	Smart structures that automatically reconfigure in-flight to optimize the range/loiter profile	Unique

engineers to believe that such developments are far from ready for application in hypersonic vehicles.

Years of investments by many agencies, including ONR's Materials by Design program, have led to the development of many computational tools that individually contribute to materials development by enabling the calculation of specific properties and their links to alloy formulation and processing. Two recent DOD programs have led to advances in integrating these isolated computational tools into systematic design tools targeted at coupling materials development and system design to reduce materials development time and cost while optimizing the property suite. DARPA led the Accelerated Insertion of Materials program, an industry-focused effort to assemble existing software into system-based design tools. The Air Force Office of Scientific Research initiated and continues to fund a program on further development of the computational tools, known as Materials Engineering for Affordable New Systems. These two programs have yet to highlight material systems of special interest to the Navy, and it would appear timely for ONR to extend its interest in computational tools in this direction, with naval aviation materials only one of the anticipated areas of emphasis. The committee notes the existence of a Multidisciplinary University Research Initiative topic in this general area intended for initiation in fiscal year 2005, but suggests that this effort should be supplemented by and integrated into a strong core effort.

Structures and Materials for Heavy-Lift Air Transport

The committee notes that, among the many other issues that will be important in the design and construction of next-generation rotorcraft/vertical short takeoff and landing (VSTOL) and dirigibles as future heavy-lift vehicles, the scale-up in size from present prototypical vessels will demand lightweight structural members, with stiffness far exceeding that of any currently available materials. Furthermore, these members are likely to be subjected to low-cycle fatigue stresses that would be expected to severely limit the lifetimes of parts fabricated from today's materials. Among the list of research options that should be considered is further attention to organic matrix composites. These materials have been the subject of broad investigation for decades, and their successful introduction into aircraft manufacture is strong testimony to the value of such long-term, sustained research and development investment programs. However, present-day materials are costly to produce and fabricate into parts, are not yet as "unobservable" as future capabilities demand, and are subject to corrosion (oxidation) damage at unacceptable rates, severely limiting lifetimes. Furthermore, when they are used as structural members, their sensitivity to fatigue fracture may be a life-limiting feature. Two important new advances in materials development promise to address these deficiencies and herald a new generation of organic matrix composites, but extensive research is required to achieve further basic understanding and apply these principles to specific systems. These two advances are the increased availability of low-cost multiwall nanotubes for use as strengthening fibers and the invention of strategies for the fabrication of self-healing fatigue-resistant materials.

Nanotubes. Carbon nanotubes offer the opportunity for control of strength and electrical conductivity in the materials used as strengthening fiber for organic matrix composites. Single-walled tubes remain expensive and difficult to process, although that may change in the near future as a result of many research endeavors worldwide. Multiwall tubes are already available in large quantities at reasonable prices and may be quite suitable for composites fabrication. Research topics should include development of nanotube processing technologies; functionalization of nanotubes for and "compatibilization" with various organic matrices; characterization of mechanical properties, including lifetime in hostile environments; characterization of physical properties, especially mechanical properties and radiation visibility; and all of the above under various conditions involving process variables.

Self-Healing Composites. Current strategies for self-healing composites are based on inclusion of microscopic encapsulated epoxies that are activated by crack propagation and polymerize to strengthen the material ahead of the moving crack tip and arrest its further motion. This proven concept needs further development, primarily in the area of further miniaturization of the encapsulants to eliminate their current composite mechanical property degradation.

Human Systems Integration

Human systems integration (HSI) optimizes the human part of the total system equation by integrating human factors engineering; manpower, personnel, and training; health hazards; safety factors; medical factors; personnel (or human) survivability factors; and habitability considerations into the systems acquisition process.[16] All of these factors are critical to naval aviation. A committee review of the briefings and supplementary materials provided to it as well as answers to posed questions identified both strengths and weaknesses in the existing program for HSI and established the basis for recommendations for extensive program augmentation.

[16]See <http://iac.dtic.mil/hsiac/HSI.htm>. Last accessed on January 21, 2005.

Human Systems Integration S&T Opportunities

The following promising naval aviation S&T opportunities in HSI have been organized by time frame, naval relevance, naval capability, and level of research and are summarized in Table 3.4. Subsequent paragraphs amplify some of the proposed areas of research.

Human Systems Integration Research in the 2007 to 2010 Time Frame

Naval aviation is unique in its requirement for maintaining situational awareness from seabed to space. To provide this capability, basic research in machine vision is essential for detecting and identifying entities from seabed to space. No human could maintain vigilance over this large domain. No human could detect entities at a reliably rapid rate. Given the mandate to reduce the number of personnel on aircraft carriers, dependence on machines for detecting and identifying entities is mandatory. A consortium of 220 organizations has developed an online machine vision information exchange.[17] This would be an excellent place for ONR to leverage technologies and to identify category 6.1 research that is naval unique.

Next in consideration should be the tagging, tracking, and locating of space, air, surface, and subsurface entities to maintain total battle space awareness. The Air Force Research Laboratory (AFRL) has a long history of developing technologies for tracking air and ground targets.[18] ONR's collaboration with the AFRL to extend these techniques to surface and subsurface targets would dramatically move the Navy to total battle space awareness.

One part of battle space awareness that is consistently lacking is knowledge of own-force performance, including command and control performance. ONR is currently funding research in this area.[19] This work could be developed into an automatic performance measurement system for shipboard and airborne command and control.

While the above S&T opportunities address naval-unique needs, the following address naval-essential needs. Basic research in information fusion is required by all three aviation Services to provide rapid and accurate decision support. ONR can use and leverage research reported by the International Society of Information Fusion.[20] This contact will also decrease the risk of technological surprise in this critical technology area.

---

[17]See <http://www.machinevisiononline.org>. Last accessed on January 21, 2005.

[18]See <http://www.afrlhorizons.com/Briefs/Feb04/SN0308.html>. Last accessed on January 21, 2005.

[19]See Kathleen Carley, Carnegie Mellon University, "Typology of C2 Performance," <http://www.casos.cs.cmu.edu/events/conferences/template.doc>. Last accessed on January 21, 2005.

[20]See <http://www.inforfusion.org>. Last accessed on January 21, 2005.

In a more applied area, the AFRL released a grant notice in May 2004 for Revolutionary Automatic Target Recognition and Sensor Research.[21] ONR collaboration with this grant process could leverage technologies for rapid and accurate decision support in naval aviation.

S&T opportunities for the 2007 to 2010 time frame that are naval relevant should be focused on interoperability. The least-understood interoperability issue is fitting a human into multiple roles. What are the dimensions of human interoperability? Is there an underlying skill that makes some sailors and aviators better able to perform various roles—pilot, mission planner, instructor, inspector, and so on? Are there human skills that are more interoperable than others? Are there aspects of tasks that enhance human interoperability? Are there organizational structures that enhance human interoperability? "Human interoperability" is not a new term; it was used in 2001 by then-Secretary General of NATO Lord Robertson in a now-famous speech.[22] Again, this research is important not only for enhancing capability for naval aviation but also for avoiding technology surprise.

By identifying ways to select and train personnel for multirole tasks, applied research also addresses the requirement for reduced manning of ships. In addition to the ability to handle multiple roles with efficiency and accuracy, there is the ability to adapt. Given the reduced manning of naval aircraft carriers, designers will not be able to provide solutions for every contingency, or for every combination of personnel and role. Therefore, selecting and training personnel for adaptability will be a key element of mission success. Soldier adaptability is a critical component of Army transformation.[23] Again, ONR can leverage another Service's research and apply it to naval aviation.

Human Systems Integration Research in the 2011 to 2025 Time Frame

To provide minimally manned naval air operations, ONR must perform basic research in the synergistic teaming of manned and unmanned assets. The tradeoffs of various levels of synergy have already been identified by the Army.[24] The Aviation Applied Technology Directorate has listed its desired capabilities for rotary-wing aircraft and UAVs. The research to best understand the potentials for

[21]See <http://www.fedgrants.gov/Applicants/USAF/AFRL/Wright/04-03-SNK-Amendment3/Grant.html>. Last accessed on January 21, 2005.

[22]See <http://www.nato.int/docu/speech/2001/s010928a.htm>. Last accessed on January 21, 2005.

[23]See <http://www.ausa.org/transformation/article_milestones.html>. Last accessed on January 21, 2005.

[24]See, for example, <http://www.redstone.army.mil/pub_affairs/archive/2003/04Apr2003/articles/0423103140801.html>. Last accessed on January 21, 2005.

TABLE 3.4 Human Systems Integration

Time Frame	Naval Relevance	Naval Capability	Basic Research (6.1)	Applied Research (6.2)	Advanced Technology Development (6.3)
2007-2010	Naval unique	Total battle space awareness	Machine vision	Tagging, tracking, and locating space, air, surface, and subsurface entities	Automatic performance measurement, assessment, and feedback; measures of command and control performance
2007-2010	Naval essential	Rapid and accurate decision support	Information fusion	Automatic target recognition	Digital simulation of hanger bays to support operations research
2007-2010	Naval relevant	Interoperability	Human interoperability	Selection and training for multi-role sailors and aviators; selecting personnel for multitasking and adaptability	Compatibility among assets (i.e., ships, submarines, aircraft, ground vehicles)

2011-2025	Naval unique	Minimally manned naval air operations	Synergistic teaming of manned and unmanned assets	Semiautonomous fleets of unmanned aerial vehicles, unmanned underwater vehicles, unmanned ground vehicles	Automation for reducing or eliminating support personnel; autonomous carrier protection
2011-2025	Naval essential	24/7/365 combat readiness	Computer aiding	Tactical fatigue reduction, immersion protection, full-spectrum laser eye protection, small footprint sea-based and reach-back-enabled training and mission rehearsal tools	Spinal injury mitigation for both pilots and operators of high-speed boats, fuel handling protection, multilevel security for training and mission rehearsal tools, semiautomated forces behavior models for training simulators
2011-2025	Naval relevant	Intuitive information assimilation	Methods for presenting uncertainty and assessing validity of data	Holographic displays, verbal command for mission-planning support	Handheld and wearable displays

synergies and the risks of incompatibilities must be performed if naval aviation is to be conducted with mixed fleets of manned and unmanned assets.

To minimize the number of personnel on aircraft carriers, routine (deck scrubbing) and hazardous (patrol) jobs could be performed by fleets of semi-autonomous unmanned ground vehicles (for deck scrubbing), unmanned underwater vehicles (for underwater patrol), unattended ground sensors (for beach and harbor patrol), and UAVs (for ship, beach, and harbor patrol). The Space and Naval Warfare Systems Command has performed extensive research in this area already.[25] Again, ONR could leverage this research for naval aviation.

Identifying automation technologies for reducing or eliminating support personnel and providing autonomous carrier protection that could transition to the fleet in the 2011 to 2025 time frame would greatly reduce manning requirements. Such a capability is already listed in the Navy Functional Concept of Operations.[26] Autonomous aircraft carrier protection has been identified as a requirement by the Chief of Naval Operations.[27]

A naval-essential capability in the 2011 to 2025 time frame is 24/7/365 combat readiness. With limited assets, the naval aviation community must have computer aiding. Computer aiding includes the support of human operators in all aspects of operations. Some basic research has been performed in adaptive automation in which a computer performs more tasks as operators/aviators become overloaded. The Office of the Secretary of Defense has funded work on adaptive automation for supervisory control of UAVs.[28] Supporting operators/aviators physically has not been addressed, nor has cognitive aiding. Does the operator/aviator need enhanced perception, information processing, or response support? These questions have not been adequately addressed.

Given the need to provide 24/7/365 combat readiness, human operators must be used to their maximum efficient endurance. Methods must be developed for tactical fatigue reduction. Use of go/no-go drugs, sleep/rest schedules, and so on has been explored by the National Aeronautics and Space Administration (NASA)[29] and the AFRL.[30] These topics both present excellent collaboration opportunities for ONR, given the Navy's unique seaboard accommodations. Also

---

[25]See <http://enterprise.spawar.navy.mil/body.cfm?Topic_ID=699&Type=R&category=22&subcat=31>. Last accessed on January 21, 2005.

[26]See <http://www.dmdc.osd.mil/jrio/documents/Navy_Functional_CONOPS_v1-1.pdf>. Last accessed on January 21, 2005.

[27]See <http://nationaldefense.ndia.org/article.cfm?Id=1356>. Last accessed on January 21, 2005.

[28]See <http://www.dtic.mil/matris/sbir/sbir041/srch/sbir370.html>. Last accessed on January 21, 2005.

[29]See <http://www.hq.nasa.gov/office/legaff/mann8-3.html>. Last accessed on January 21, 2005.

[30]See <http://www.afrlhorizons.com/Briefs/Jun03/HE0301.html>. Last accessed on January 21, 2005.

critical to providing 24/7/365 combat readiness is reducing the risk of loss of naval aviators. This includes immersion protection for downed pilots and full-spectrum laser eye protection. Finally, maintaining combat readiness requires just-in-time training, which of necessity must be provided shipboard through embedded training, ruggedized sea-based trainers, and reach-back trainers. These trainers could also provide mission rehearsal.

Similarly, S&T 6.3 opportunities are focused on reducing risk to naval aviation personnel. These opportunities include spinal cord injury mitigation for both pilots and operators of high-speed boats, fuel-handling protection, multilevel security for training and mission rehearsal tools, and semi-automated forces behavioral models for training simulators. There is a 40-year history of research on spinal cord injury mitigation.[31] Fuel-handling technologies have been developed by the Air Force[32] and the Army Corps of Engineers.[33] The Navy has a long history of interest in mission rehearsal,[34] and ties between the Naval Aviation Training System Program Manager, PMA205, and ONR would enhance naval capabilities in this critical area. Finally, the Army has performed extensive research in semi-automated forces for training. The Army Program Executive Office for Simulation, Training, and Instrumentation has models that could be applied directly to naval aviation trainers.[35]

Naval-relevant research in the 2011 to 2025 time frame also includes basic research in command and control and piloting that makes assimilation of information intuitive. The Space and Naval Warfare Systems Command presented its research at the 2004 Command and Control Research and Technology Symposium.[36] Such a symposium would provide ONR with collaborators as well as technologies to leverage and would reduce technology surprise.

Finally, 6.3 S&T opportunities for the 2011 to 2025 time frame include the use of handheld and wearable displays. Such displays have already been developed for F-15 maintainers.[37] One of the most time-intensive tasks aboard ship is maintaining the technical orders for aircraft and aircraft ground support equip-

---

[31]See <https://fhp.osd.mil/congress/pdfs/index.pdf>. Last accessed on January 21, 2005.

[32]See <http://www2.acc.af.mil/combat-edge/past_issues/July%202004/Stories/0704story2.htm>. Last accessed on January 21, 2005.

[33]See <http://www.swf.usace.army.mil/pubdata/ed/elect/SPECFAC.asp>. Last accessed on January 21, 2005.

[34]See <http://pma205.navair.navy.mil/briefs/mrwgjuly98/>. Last accessed on January 21, 2005.

[35]See <http://www.peostri.army.mil/PRODUCTS/MODSAF-PMITTS>. Last accessed on January 21, 2005.

[36]See <http://www.dodccrp.org/events/2004/CCRTS_San_Diego/CD/papers/063.pdf>. Last accessed on January 21, 2005.

[37]See <http://www.darpa.mil/mto/smartmod/presentation/factsheets/maintenance.html>. Last accessed on January 21, 2005.

ment. What if all technical orders were electronic, and what if only the relevant portions were presented during a maintenance task?

Survivability

Survivability is achieved according to a multidiscipline approach involving avionics, sensors, propulsion/power, structures and materials, and so on. These areas must be worked on collectively as an integrated product to arrive at an affordable and survivable design. The ONR presentations reviewed by the committee did not address survivability either as a functional requirement of a particular ONR program or as a functional discipline within the ONR S&T program community.

Survivability is a committee concern in five areas, as indicated in Table 3.1:

- High-altitude, long-endurance UAVs will have a significant IR contrast signature due to the cold background.
- Tactical, high-performance UAVs will encounter similar LO challenges as manned tactical aircraft.
- Air-breathing hypersonic vehicles will have a significant IR signature due to their speed.
- Next-generation rotorcraft/VSTOL air vehicles will have unique rotor signatures.
- Heavy-lift vehicles will be large, low flyers with visual signature challenges.

The committee recalls that ONR used to have an overarching survivability systems engineering department that was chartered with defining the state of the art in terms of survivability (susceptibility, vulnerability, and recoverability) and developing a D&I road map to guide ONR's activities. Coordination with other Services was part of the effort, since there exists a large database of technologies and programs that specifically address survivability concerns.

Core Technologies

In addition to the aviation technologies discussed above, ONR must devote S&T efforts to the fundamental core technologies of aerodynamics and dynamic modeling and simulation. Two important application areas for these core technologies are discussed here: heavy-lift vehicles and UAVs.

A new DOD heavy-lift rotorcraft (currently designated as Air Maneuver Transport, and Air Theater Transport) in the 13- to 20-ton weight range with a capability to carry 50 to 60 passengers is under consideration for development by the Joint Services (Army, Navy) and NASA. There are ongoing efforts, with participation from contractors, to conduct studies of trade-offs to help evolve and define the system's requirements. The heavy-lift requirements (e.g., cruise speed, maximum dash speed, maximum payload, endurance, survivability), design considerations (dynamic system, rotor blades, flight control system, avionics, power plant, including engine), and operational envelope are still evolving and are not known to this committee.

The committee envisions that the new heavy-lift rotorcraft program will leverage many of the existing analytical tools in both government and industry in the development process. The committee also believes that it is critical that the naval-unique requirements be identified and defined up front as part of the overall joint services requirements to mitigate trade-off and design risks. Such a vehicle is critically important to the Sea Basing pillar of Sea Power 21 and to the Marine Corps in its vision of attacking from the sea. There will be many new S&T challenges in the development of this heavy-lift vehicle, and ONR should become engaged from the beginning in creating S&T solutions to the challenges.

There is a need to study aerodynamic issues for micro-UAVs. In some new concepts, propulsion and aerodynamics may become more strongly integrated. At the smaller sizes and lower speeds for these vehicles, the aerodynamics could differ significantly because the lower Reynolds number might easily be in a transition range. This could mean lower lift-to-drag ratios that adversely affect range. Microelectromechanical system technology might be used on lifting surfaces to provide triggers for inducing turbulence so as to reduce drag due to flow separation. Mechanisms for lift production might be completely re-examined; for example, unsteady aircraft motion could be used to produce lift plus possibly some propulsive thrust. The wing flapping of birds is just one example of how unsteady motion of air-vehicle components produces both lift and thrust. Again, scale-down of existing technologies probably would not be optimal, which implies a great need for research and technology development.

ONR should consider the dynamics of small UAVs that maneuver sharply (high forward and turning accelerations) to serve on missions that go beyond loitering at a distance from objects of interest that require proximity with survivability. The time scales for the dynamics of these aircraft decrease as the size decreases. Therefore, remote control introduces problems at the operator-machine interface. New types of automatic controls may be required to handle the fine aspects of flight while the remote pilot maintains general control. New propulsive schemes and new aerodynamic designs would be needed to enable the bursts of acceleration and sharp turns that protect these aircraft from destruction as they fly near hostile forces.

Beyond these specific problem areas, aerodynamics should be considered a core technology area to be sustained by ONR. For example, it does have substantial importance in hypersonic flight technology and as a key enabling technology for understanding and exploitation of dynamic fluid-structure interaction. The state of the art for time-dependent or unsteady aerodynamics is ripe for rapid advancement and that, in turn, will enable new advances in the understanding of and better design for improved life of structural components for high-cycle fatigue, rapid maneuvers both desirable and hazardous (abrupt wing stall), and quieter aircraft and rotorcraft (reduced acoustic signatures).

Modeling of flow in engine combustors can also be in a similar range. Current capabilities for modeling fluid dynamics in the Reynolds number range below fully developed turbulence but above the laminar range are not reliable. Therefore, computational fluid dynamics (CFD) capabilities must be extended to handle unsteady transitioning flows.

Noise and vibration of rotorcraft can lead to detection by an enemy force and also to fatigue of the rotorcraft crew and structure. Development of the capability to create designs with dramatically reduced noise and vibration requires fundamental improvements in dynamic modeling and simulation. There is a pressing need for accurate and computationally tractable models of the unsteady flow field and its interaction with nonlinear structural elements, including rotor blades.

Similar fundamental issues arise in the design of turbo-machinery blades for long life in a harsh flow environment that too often leads to high-cycle fatigue of these blades.

The complex aerodynamic flow field in which aircraft and rotorcraft take off and land from an aircraft carrier or other ship can give rise to dynamic stability and control phenomena that also require fundamental advances in the understanding of unsteady aerodynamic flows to predict the motion and stability of flight vehicles in the shipboard environment.

Abrupt wing stall (AWS) is a dynamic phenomenon that is characterized by the feedback coupling between a complex (separated) flow field with the motion of the vehicle. Typically AWS occurs at high angles of attack and high subsonic/transonic Mach numbers. Several currently operational vehicles, including the F/A-18, have experienced AWS during their development program. Current CFD models and simulation methods are inadequate to simulate AWS, and successful correlation between computational simulations and experiment (wind tunnel or flight tests) has not yet been fully achieved. Fundamental improvements in both understanding of the underlying physical phenomena and computational algorithms for efficient simulation are required.

In practice, each of the topics discussed above is often treated separately by largely distinct communities of engineers and scientists. Such compartmentalization tends to emphasize the differences and idiosyncrasies among these several topics. But fundamentally the underlying phenomena have much in common, and a vigorous D&I program in unsteady aerodynamics and fluid-structure interaction could pay rich dividends in the capability to model and simulate, and thereby reduce and ameliorate, rotorcraft noise, turbo-machinery high-cycle fatigue, and AWS.

FINDINGS

- For the disruptive capabilities identified as highly desirable for Naval Power 21, the committee has identified the technologies requiring a strong, well-planned, focused effort to achieve an order-of-magnitude improvement. These can be categorized under the functional areas of avionics, sensors, propulsion and power, structures and materials, human factors, survivability, and the core elements of aerodynamics and dynamic modeling and simulation.
- ONR is no longer pursuing leading-edge survivability S&T, nor is it connected to the other Services and their efforts in this area.

RECOMMENDATION

- ONR should create an overarching survivability systems engineering effort that is chartered with defining the state of the art in terms of survivability (susceptibility, vulnerability, and recoverability) and developing a D&I road map to guide naval S&T activities. ONR should consider developing strong technology programs supporting the desired disruptive capabilities if these are to become a reality.

4

Science and Technology Planning for Naval Aviation

To evaluate the S&T planning processes that support technology development at ONR, the committee received briefings from ONR as well as from the Army and the Air Force. Two very different planning approaches were evident. ONR takes a heuristic approach and organizes its projects by portfolio—an approach that makes it difficult to link to a process for meeting identified requirements or to project deliverables and that also presents a management challenge to execute in an integrated format responsive to identified capability needs and gaps. In contrast, the Air Force Research Laboratory (AFRL) takes a systems engineering approach and organizes its efforts using the Integrated Product and Process Development (IPPD) method. The systems engineering approach is much easier than the heuristics approach to manage and to integrate with other S&T efforts.

ONR—THE PORTFOLIO APPROACH

Although Naval Studies Board committees in 2001 and 2002 recommended that ONR adopt a systems engineering approach to technology development,[1] ONR continues with a portfolio of technically oriented projects. ONR's S&T activities related to a systems-level area are not concentrated in a single organiza-

[1] See Naval Studies Board, National Research Council, 2001, *2001 Assessment of the Office of Naval Research's Aircraft Technology Program*, National Academy Press, Washington, D.C.; and Naval Studies Board, National Research Council, 2002, *2002 Assessment of the Office of Naval Research's Air and Surface Weapons Technology Program*, National Academy Press, Washington, D.C.

tion but rather are conducted within several departments. ONR is not organizationally structured along war-fighting functional areas, such as naval aviation, surface ship warfare, or weapons systems, but according to technical discipline such as electronics, materials, or human systems. Although such a structure is not uncommon in S&T organizations, it is often complemented by a program office structure representing war-fighting discipline areas and led by individuals who are responsible for the funding and management of technology development across many organizational and technical disciplines. Without such a matrix approach, it is difficult to achieve an integrated and efficient technical program. And that is true for ONR today.

ONR's portfolio of naval aviation S&T appears to be both diffuse and eclectic. Because no single individual or organization was responsible for the ensemble, ONR could not provide the committee with a coherent naval aviation S&T program plan across the entire organization. ONR staff did give briefings on a few selective thrusts that seemed well balanced and well supported and would definitely benefit many naval aviation elements.

AIR FORCE—THE INTEGRATED PRODUCT AND PROCESS DEVELOPMENT APPROACH

The Department of Defense (DOD) accepted IPPD as a preferred systems engineering process in 1996. IPPD activities focus on the customer and meeting the customer's needs. In selecting a new technology for development, trade-offs between competing approaches are analyzed, with consideration given to operational, design, performance, production, support, and cost factors to optimize the ultimate application or system (product or service) over its life cycle.

The Air Force has embraced IPPD, and AFRL uses it as the preferred process for its S&T development. The AFRL approach (Figure 4.1) is a six-step process involving (1) determining requirements, (2) establishing S&T exit criteria (for a program to leave S&T as successful or unsuccessful), (3) developing technology alternatives, (4) performing value analyses, (5) developing and demonstrating new technologies, and (6) analyzing and delivering results.[2]

The objective of the IPPD approach is to better quantify the costs and risks associated with new technologies during the later stages of S&T development and to reduce the costs and risks of transitioning new technologies to weapons systems. The aviation S&T program at AFRL is well structured, vetted across all stakeholders, linked to capabilities, and geared to deliverables.

---

[2]The AFRL approach is described at <http://www.jgai.com/stprocess50/pm60draft/index.htm>. Last accessed on March 15, 2005.

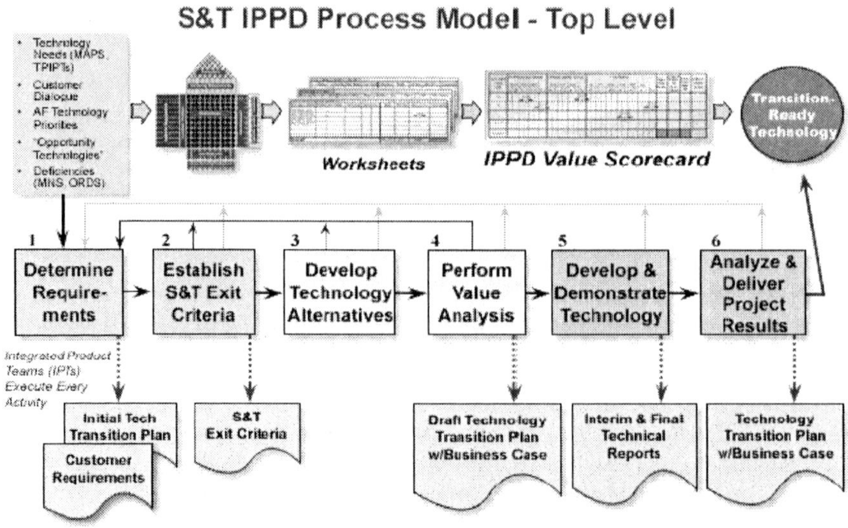

FIGURE 4.1 Integrated Product and Process Development (IPPD): Part of the AFRL culture. The AFRL IPPD approach to S&T development. SOURCE: Air Force Research Laboratory.

USING CONTEXT IN MANAGING S&T PLANNING

Although the Navy has identified IPPD as a preferred process for program managers,[3] ONR has not yet embraced it for S&T planning. In fact, as noted above, the committee found no system at ONR for assessing the overall needs for and balance in the naval aviation S&T portfolio.

Managing S&T through a portfolio is difficult, but it can be made easier by keeping in mind the larger context in which S&T initiatives are proposed. Judging which initiatives should be fostered in a portfolio depends on their contributions to the larger goals identified for naval aviation. Absent a systems engineering approach, the challenge for ONR is to discern the goals that naval aviation is intended to fulfill, identify which S&T initiatives would further their achievement, and then manage those S&T activities accordingly. This calls for a host of skills, including judging scientific merit, identifying likely impacts, estimating costs and risks, and evaluating new S&T opportunities vis-à-vis trends in naval aviation. Trends that impinge on naval aviation can be useful as benchmarks against which to evaluate proposed characteristics and new performance levels. The committee lists nine trends evident in the larger context that can be used by ONR as guidelines in its S&T planning for naval aviation:

---

[3]See <http://www.abm.rda.hq.navy.mil/navyaos>. Last accessed on March 15, 2005.

- *Substitution of technology manpower.* Technology is replacing manpower—to reduce risks to and the exposure of personnel and to minimize the costs that accompany the use of people. This trend will accelerate. A recent manifestation is the rapid introduction of unmanned aircraft, which has raised important operational problems of integrating manned and unmanned aircraft into the congested space of shipboard flight operations—both on deck and in the nearby airspace. Problems related to safety and efficiency are evident, but solutions that maintain the operational tempo achieved by manned aircraft are not.
- *Vertical launch and recovery.* Fixed-wing aircraft capable of vertical launch and recovery are now used at naval bases and aboard ships. Providing vertical launch capability means sacrificing some at-target capabilities in the aircraft, a trade-off that not only calls for a complex balancing of design characteristics and performance but also requires changes in the bases and ships that operate the aircraft.
- *Constrained operating space.* Aircraft operations from ships at sea have always been difficult due to the severely constrained operating space on ships, an issue that will become more acute with the potential integration of unmanned aircraft with manned aircraft on the same ships.
- *Information management.* The growing demand for information generation and management by U.S. forces sent to a trouble spot calls for dwell time, sophisticated use of sensors, and on-scene interpretation of disparate and sometimes conflicting data. Networked information is now a mandatory system capability.
- *More precise strike.* Very precise strike minimizes collateral damage and provides deft application of force; very accurate weaponry can thereby shorten campaigns or otherwise enable their effective management.
- *Greater capability/greater cost.* The ever greater capability built into each platform renders it more costly. When overall spending for Navy platforms is held constant, a result is fewer platforms in inventory. Spreading development and support costs by broadening the number of participants can reduce research and development costs as a fraction of total lifetime costs, freeing marginal dollars for procurement of extra aircraft.
- *Technologies giving rise to countertechnologies.* The development of counter-stealth capabilities in radars and missiles, and of counter-countermeasures in electronic warfare, makes clear that S&T management must account for possible disruptions or even a reversal of a trend.
- *Fewer types, models, and series.* The number of different naval aviation aircraft types has been declining, with consequent streamlining of maintenance, logistics, and training. This growing commonality provides opportunities for high-payoff S&T if it is focused correctly on the fewer aircraft types. Similarly, aviation weaponry and information management are increasingly standardized in the Navy and the Marine Corps and even across DOD.

- *Operational integration.* One key warfare mission area that benefits from the embarkation of Marine Corps squadrons on carriers as part of Navy carrier air wings is amphibious warfare, in which at-sea forces support Marines landing ashore. Sea basing for landed Marines presents new opportunities for S&T to enable successful implementation of this concept.

MANAGING S&T USING THE PORTFOLIO APPROACH

While the committee recognizes that managing S&T within a portfolio enviroment is difficult, it can be done. However, the S&T communities must be active participants in identifying the key technology areas, and S&T performers and users must be organized into integrated product develoment teams representing capability areas and encouraged to explore the range of technology options available to achieve the particular capability. Naval aviation war fighters and systems planners should be charged with giving technologists "I wish I could . . ." information. Technology experts and visionaries should be asking users: "If you had the following technology to solve your capability gap, would you use it?" Effective planning can be achieved by bringing these two communities together, aided by system studies, contributions from industry, input from the scientific community via program managers at ONR and scientific experts from the Service laboratories, and outreach to other agencies and private-sector expertise. The list of needed technologies must be strongly coupled to the users' list of desired capabilities.

Technology options to address capabilities may shift in time, but they do form the basis for S&T planning. A difficult part of this process is to group technologies according to near-term, mid-term, and long-term needs. A second and perhaps more difficult part is to ensure that adequate funding and a management structure are present to achieve cross-technology, integrated solutions. A third difficult, but doable, problem relates to the truly revolutionary breakthrough technology solutions that cannot be planned. These will arise from the inventiveness of scientists and engineers, some of whom may be funded by ONR and others not. S&T planning should have the flexibility to ensure ONR's participation in supporting these creative endeavors, and ONR's execution plan should provide for incorporation of these game-changing options into future-year versions of the evolving S&T plan.

When the resource base is finite, prioritization of technology options requires system studies; sensitivity to the need for cross-Service, cross-agency cooperation; and recognition of the DOD reliance process.[4] Each of the time

---

[4]Under the DOD Reliance programs, Service S&T portfolios are reviewed regularly by the DDR&E and the Services are urged (and directed) not to unnecessarily duplicate efforts, but to depend instead on the results obtained in another Service's S&T program's agreed-on areas.

frames demands different prioritization processes. For example, the process for developing Future Naval Capabilities (FNCs) properly links the 6.3 agenda to analysis scenarios and war-fighter priorities. The Discovery and Invention (D&I) agenda, however, must be guided strongly by science opportunities and the perceived relevance to key technology needs.

Reducing the tendency toward stovepipe solutions is the most complex aspect of any S&T organization. In the context of this discussion, ONR has significant advantages in managing the entire portfolio of 6.1 to 6.3 naval aviation S&T efforts. But to properly exploit that advantage, ONR must balance its efforts across disciplines, ensure that resources are moved when technology transitions, and ensure that the naval laboratories remain strong participants in the overall effort.

MANAGING PROGRAMS IN AN S&T PORTFOLIO

For most or all of the near-term naval aviation technology needs, 6.3 programs are already in place with road maps, and the primary issues to be resolved are their technology readiness, funding gaps, and technology transition paths. In the Navy it appears that most of these programs fall in the FNC categories, and the planning and process for prioritization have been developed. Milestones linked to technology readiness levels and a carefully monitored process of review and transition agreements are critical management tools.

For the Navy, as for most S&T organizations, mid-term technologies will be the most difficult to plan for. The funding divisions of Discovery and Invention (6.1 and early 6.2) and Exploitation and Development (late 6.2 and 6.3) make it unclear how mid-term needs can be addressed.

Long-term requirements should be addressed by primarily 6.1 efforts, which should reflect the need for both capability-linked programs and novel concepts or breakthrough technologies. These programs do not lend themselves to rigid milestones, and allocation of resources has to be based on opportunity, scientific judgment, and a vision of potential applicability to naval aviation. Especially for 6.1 efforts, the scientific community must be brought to the table and organized to identify the critical areas for funding. For example, when the Air Force Office of Scientific Research examined the capabilities required for joint, network-centric collaboration with coalition partners, the involved 6.1 scientific community helped to identify team training as an area requiring radical new methodology.

At every stage of planning and execution, cognizance of jointness within DOD is imperative. ONR S&T planning teams must draw from the experience of the other Services, the National Aeronautics and Space Administration (NASA), and the Defense Advanced Research Projects Agency, using their planning expertise as input for prioritization of the portfolio and distribution of the resources within the Navy S&T program. In the execution of programs, collaboration with others can and does take all forms.

PROGRAM EXECUTION IN AN S&T PORTFOLIO

The goal in program execution should be to obtain the most creative solutions possible from the best performers, which implies open competition. The committee believes that ONR funding is awarded competitively, with the exception of the block funding for the Naval Research Laboratory (NRL) and the congressionally mandated programs that ONR typically executes. Integration of open-competition funding into the naval aviation S&T plan would be a natural by-product of developing the systems engineering approach described above and recommended in earlier reports.[5]

ONR should build in periodic internal and external review of its entire S&T plan, organized as a comprehensive review across all ONR codes and including the core NRL programs. Criteria for these reviews should include the rationale for program planning, funding levels for road-mapped programs, technology readiness matched to program plan, quality of research work measured against world-class standards, technology transition plans, and execution.

Providing support for discovery of novel concepts, breakthrough technologies, and improvements in the technology base is a 6.1 activity. Success requires program managers who are current in technology research and development (including the visionary scientists at laboratories such as NRL, AFRL, the Army Research Laboratory, at NASA, in industry, and so on) and at the same time cognizant of the naval aviation S&T plan. ONR must maintain currency in nanotechnology, biology, spintronics, and other areas to be ready for solutions as yet undiscovered. The determination of how much funding and what criteria to use for allotting support must be a part of the naval aviation strategic S&T plan.

A NAVAL AVIATION STRATEGIC S&T PLAN

In discussions with the committee, ONR and the Naval Air Systems Command (NAVAIR) readily agreed that they did not have a strategic S&T planning process to support naval aviation. They agreed that one was needed and that they would work together to produce it. Without a plan to assess, the committee offers

[5]See Naval Studies Board, National Research Council, 2001, *2001 Assessment of the Office of Naval Research's Aircraft Technology Program*, National Academy Press, Washington, D.C.; and Naval Studies Board, National Research Council, 2002, *2002 Assessment of the Office of Naval Research's Air and Surface Weapons Technology Program*, National Academy Press, Washington, D.C.

the following thoughts, which may be helpful to ONR and NAVAIR as they put together the first naval aviation strategic S&T plan.

A naval aviation strategic S&T plan will encompass an overall picture of naval aviation, from operations to S&T, and of the trends that impinge on naval aviation, such as those listed above by the committee. It will include identification of specific goals and development of implementation plans to reach those goals. The Office of the Chief of Naval Operations (OPNAV) has the primary responsibility for creating a naval aviation strategic plan. OPNAV will work with the Commander of Naval Air Forces (CNAF), NAVAIR, and ONR. The methodology currently being used at OPNAV to develop a strategic approach to near-term naval war fighting is the Naval Capabilities Development Process (NCDP): A range of likely future war-fighting campaigns, including joint campaigns with other U.S. military and allied forces, is analyzed to derive a measure of the current state of naval capabilities. Gaps in desired capabilities are identified, and plans to correct or fill those gaps are created. New technology insertion is one way to correct the gaps, and indeed the outputs of the FNCs are being focused on that objective.

It is essential that the Navy and the Marine Corps use processes like the NCDP and apply them to the new war-fighting operations envisioned in Naval Power 21. The Strategic Studies Group XXIII, which reports directly to the Chief of Naval Operations and has responsibility for generating revolutionary concepts for future naval war fighting, has recognized this process need and (in its report entitled *Global Maritime Fight . . . 2030 and Beyond*) is recommending a future concept development group separate from that for current fleet operations.[6] The Naval Aviation Enterprise (CNAF, OPNAV, and NAVAIR) has just created the document *Naval Aviation Vision 2020,* which can provide an initial basis for identifying future needs.[7] ONR, working together with OPNAVand NAVAIR, should be able to identify strategic goals that relate to desired operational capabilities. Once these strategic goals are established, ONR, working with NAVAIR, can create a strategic S&T plan to satisfy the technical requirements of the desired capabilities. The goals and approach of the naval aviation strategic S&T plan should be agreed to by all of the principal stakeholders in naval aviation. Following approval of the plan, ONR should be held accountable, through periodic reviews, for progress toward accomplishing the goals.

---

[6]ADM James R. Hogg, USN (Ret.), Director, CNO Strategic Studies Group, personal communication, August 10, 2005.

[7]See VADM James M. Zoortman, USN, Commander, Naval Air Forces; VADM Walter B. Massenburg, USN, Commander, Naval Air Systems Command; and RDML Thomas J. Kilcline, Jr., USN, Director, Air Warfare Division, 2005, *Naval Aviation Vision 2020*, Naval Aviation Enterprise, Department of the Navy, Washington, D.C. Available online at <http://www.nae.cnaf.navy.mil/demo/main.asp?ItemID=12>. Last accessed on September 30, 2005.

The Chief of Naval Research must drive the strategic planning at ONR. He controls the senior staff, who understand the ONR mission, the challenges and obstacles that ONR faces, and the resources available to ONR to accomplish the mission. Supporting the senior staff should be a strategic planning process manager (traffic cop) to keep the activities moving toward completion of the strategic plan. These activities will incorporate both near-term and far-term elements. Near-term needs are addressed by FNC activities and far-term needs by D&I activities.

An effective strategic plan is developed by those who will implement it and coordinated by those who will use it.[8] In the case of naval aviation S&T, this means that close coordination among ONR, NAVAIR, and OPNAV will be required to create the naval aviation strategic S&T plan. Building the plan should simultaneously incorporate opportunities that can be matured ("technology push") and crosscutting initiatives that are new ("requirements pull"). The plan must have metrics by which to measure success (e.g., technology transitions) and regular milestones that can be updated, and it should be keyed to the annual fiscal cycle.

Since developing a naval aviation strategic S&T plan will be a new endeavor for ONR, the committee offers this outline as a guide:

1. Start with an enterprise view.
    a. What are the "alternative futures" to which the vision of the future is directed?
    b. What is the naval aviation vision—what does the future hold?
    c. What is new this year versus what was said last year?
    d. What naval aviation capabilities and gaps exist within the Future Years Defense Plan (FYDP)?
    e. What new capabilities are needed in the future (outside the FYDP)?
2. Highlight and track S&T gaps and opportunities in naval aviation.
    a. Understand that ONR must become involved with the gap assessment process (get S&T gaps annually from N70, Warfare Integration).
    b. Prepare an S&T gap plan and associated deliverables.
    c. Prepare an S&T opportunities plan and associated deliverables.
    d. Coordinate gap and opportunities plans with the Air Force, Army, and NASA.
    e. Coordinate the gap and opportunities plans with NAVAIR and other system commands.

[8]The committee recognizes that ONR does not perform the actual S&T but instead funds universities, laboratories, and companies to accomplish that work. ONR's very important role is to foster the right S&T to achieve the technical goals that would enable the strategic goals and achieve the vision of naval aviation. Users will always drive a strategic plan, and for ONR the users are the fleet and the systems commands that provide capabilities to the fleet.

3. Identify the technology areas of interest to naval aviation.
   a. Do a portfolio analysis that shows investment in each technology area.
   b. Establish metrics for each: projects transitioned, stopped/divested, ongoing.
   c. Look at last year's strategic plan and metrics for success.
   d. Identify what was projected/said and what was actually achieved.
4. Track investment levels (year after year) to support strategic plan analysis.
   a. Enter investments into a database.
   b. Tag program elements and projects so that database sorting can support analysis.
   c. Tag by warfare area, mission capability, platform category, and user.
   d. Provide ONR codes with the ability to do searches and sorts.
5. Perform an annual portfolio investment analysis and include it in the annual strategic plan.
   a. Analyze short term (FNC plus Innovative Naval Prototypes) versus far term (D&I).
   b. Track S&T investment trends for naval aviation.
   c. Analyze S&T investment for surface, subsurface, and space, and the coordination of each with naval aviation.
6. Complete a technology analysis for each area of interest to naval aviation.
   a. Evaluate current state.
   b. Determine what sharing is going on, and with whom, and what joint investments are being made.
   c. Analyze last year's, this year's, and next year's proposed investments.
   d. Categorize expectations: transitions, spin-offs, terminations.

CONGRESSIONAL ADD-ONS

The development of a naval aviation strategic S&T plan will help ONR considerably with steering congressionally directed add-ons toward projects that are recognizably beneficial to naval aviation. Congressional projects supplant the core program of naval aviation S&T, an unfortunate circumstance that has been ongoing for a number of years and needs to be stopped. An examination of the President's FY 2005 budget highlights the problem.

The President's budget proposed lower amounts in Navy 6.2/6.3 S&T program elements compared with the amounts for 2004. Table 4.1 shows that funding for the 6.2 category was reduced by 22 percent and for the 6.3 category was reduced by 33 percent when compared with FY 2004 appropriations. The 6.1 category, Basic Research, remained virtually the same as for the previous year. The reductions in the 6.2 and 6.3 categories were spread across multiple program elements, as shown in Table 4.2.

TABLE 4.1 President's FY 2005 Budget for Navy S&T Compared with FY 2004 Appropriations (millions $)

		FY 2004	FY 2005	Percentage Change
6.1	Basic Research	484	477	–1
6.2	Applied Research	723	564	–22
6.3	Advanced Technology Development	1,009	677	–33
6.4	Prototypes	2,807	2,804	0
6.5	System Development and Demonstration	6,360	8,009	26
6.6	Management Support	687	654	–5
6.7	Operational System Development	2,898	3,162	9
	TOTAL	14,968	16,347	9

SOURCE: Office of Naval Research.

When Congress appropriated the FY 2005 budget, it added $589 million to ONR's account for specific S&T projects (in constituent districts). Usually 1- to 2-year efforts, these projects are widely diverse in scope and technology. These congressional add-ons disrupt Navy S&T efforts because they are short-term, unexpected projects that have no obvious connection with Navy S&T planning. They are also a burden on ONR in terms of support (contracts, finance, legal, and program management). The committee observed that these congressional add-ons have become a way of life at ONR and that the Navy anticipates them by reducing its annual requests to Congress in the ONR area so as to increase funding in other areas. Table 4.2 indicates that congressional add-ons were expected in 13 different PEs. Table 4.3 shows the effects of congressional adds-ons to just one PE at ONR (PE0602114N, Power Projection). In FY 2003, 14 projects were added; in FY 2004, 19 projects were added; and in FY 2005, 17 projects were added—on top of the ONR base S&T programs.

Congressional add-ons are not a measure of success for ONR, but rather are a net loss of discretionary S&T budget and create a significant distraction from the main Navy mission focus with programs that rarely make their way into the fleet. If ONR develops a naval aviation strategic S&T plan, congressional staffs could better funnel constituent desires into support for this plan.

TABLE 4.2 Proposed Reductions in the President's FY 2005 Budget for Navy S&T by Program Element (millions $)

	Name	Program Element	FY 2004	FY 2005	Percentage Change
6.2	Power Projection	0602114N	143	99	–31
6.3	Power Projection	0603114N	161	92	–43
6.2	Force Protection	0602123N	113	96	–15
6.3	Force Protection	0603123N	120	82	–32
6.2	Common Picture	0602235N	95	60	–37
6.2	Warfighter Sustainment	0602236N	101	64	–36
6.3	Warfighter Sustainment	0603236N	86	61	–29
6.2	Ocean War Environment	0602425N	62	48	–23
6.2	Undersea Warfare	0602747N	77	64	–17
6.3	Undersea Warfare	0603747N	47	27	–43
6.3	Radio Frequency Systems	0603271N	62	44	–29
6.3	U.S. Marine Corps ACTD	0603640N	90	58	–35
6.3	Warfighter Experiments	0603758N	39	16	–59

NOTE: ACTD, Advanced Concept Technology Demonstration. SOURCE: Office of Naval Research.

TABLE 4.3 Congressional Add-ons in Just One Program Element (thousands $)

PE0602114N	FY 2003	FY 2004	FY 2005
Aircraft Carrier Surveillance System			2,800
High Energy Thermobaric Warhead			1,000
Hybrid Stored Energy for Weapons			1,500
Hypersonics			1,000
Integrated Personal Protection System			1,200
Millimeter/Terahertz Imaging Arrays			2,100
Mobiles on Scene Sensor A/C C4I Center			1,000
Silver Fox UAV			2,500
Smart Optical Sensor Payload			1,000
Unattended Imaging Sensor Network			1,000
Kill Assist All-Weather Targeting System		3,115	1,700
Combustion Light Gas Gun		4,203	4,200
Fireladar		1,483	1,700
Nanocomposite Warheads		2,521	2,600
WBG Semiconductors		1,384	1,700
Thermal Management of Ground Stations		5,191	6,000
GAN Microelectronics and Materials		2,967	–
Intelligence Control System for SWARM UAVs		4,203	–
Multi-INT Exploitation Systems		2,769	–
Electric Actuator Technology		1,187	–
Free Electron Laser		6,923	–
Nonlinear Dynamics—Controlled Chaos		3,362	–
Radar Infrared Imaging		2,076	–
Rocket Propulsion (IHPRPT)		989	–
Chemical Weapons Detection for UAV		1,384	–
Hybrid Lidar-Radar		1,684	–
Ultra Short Pulse Laser Micromachining	1,214	1,384	–
High-Efficiency Piezoelectric Crystals	1,668	2,076	–
Mini High-Definition Digital Camera	952	–	–
Panoramic Night-Imaging System	3,240	–	–
Pulse Detonation Engine	1,003	–	–
Advanced Multifunction Receiver System	1,625	–	–
Low-Cost Fused Remote Sensors	954	–	–
Zlost Cost SWARM UAV	2,380	–	–
Millimeter Wave Infrared Imaging	1,907	–	–
SAR for All-Weather Targeting	6,712	–	–
Real-World Immersive Imaging	1,907	–	–
Hybrid Fiber Optic Wireless System	952	–	–
Integrated Biological Warfare Technology Platform	3,813	5,092	3,500
Interrogator High-Speed Retro Reflector Communications	1,627	1,978	2,000

TABLE 4.3 Continued

PE0602114N	FY 2003	FY 2004	FY 2005
ONR S&T Program			
Strategic Sustainment	9,988	23,149	31,160
High-Speed Propulsion/Advanced Weaponry	31,919	16,884	30,034
Unmanned Vehicles	15,280	14,548	11,354
Naval, EOIR, and Sensor Technology	10,298	13,259	12,283
Electric Weapons	5,780	13,400	10,000
Strike Technology	7,300	5,417	4,000
TOTAL	110,519	142,628	137,331
Congressional add-ons	29,667	55,969	38,500
Percentage of total	27%	39%	28%

SOURCE: Office of Naval Research.

FINDINGS

- There is no strategic planning process at ONR for naval aviation S&T. Current naval aviation S&T is "opportunity" driven, rather than "requirements gap" driven. There is no process to create a vision or strategies and no apparent connection between S&T investments and future naval aviation goals. S&T plans are developed from the bottom up, and funding allocations are based on current technical needs instead of broader and longer-term Navy goals and gaps in capabilities.
- ONR does not use a systems engineering approach in the planning and execution of its technology development. As a result, projects are developed ad hoc and appear to be "opportunity" driven rather than "requirements" driven. Technology gaps are not systematically identified and thus are not well defined. Systems analysis is not used to determine technology priorities or investment strategies.
- ONR's organization according to technical discipline makes it difficult for ONR to support cross-disciplinary areas, such as naval aviation. ONR currently lacks a formal process for managing naval aviation S&T, which involves multiple disciplines and programs located in six different ONR organizations. There is currently no single program manager with authority to approve a budget and long-term planning/direction setting for naval aviation S&T across ONR.
- The committee believes that the large number of congressionally directed aviation projects at ONR is counterproductive to ONR's naval aviation S&T efforts. These projects supplant the budget for core S&T efforts, add to the workload of administrators and managers, and distort planning with the introduc-

tion of short-term, unexpected projects that rarely transition into future naval capabilities. The committee views current congressional add-ons not as a measure of success for ONR, but rather as a burden and a distortion of good S&T practice.

• The committee observed poor communication among organizations—particularly between ONR and NAVAIR. Some personal relationships were good, but in general the organizational relationships were weak. Naval aviation visionaries were difficult to identify. Technologists were not interacting with war fighters or Pentagon Planning, Programming, Budgeting, and Execution System (PPBES) planners, and there was insufficient interaction and coordination with Army and Air Force aviation strategies.

RECOMMENDATIONS

• ONR should support OPNAV in its efforts to develop a naval aviation strategic S&T plan that defines the future of naval aviation and the role of naval aviation in Naval Power 21. The plan should clearly identify future technology needs, and ONR should respond to these needs by creating a naval aviation strategic S&T plan. The Chief of Naval Research (CNR) should lead development of the plan, which should be structured by technology. NAVAIR and the program executive offices should be partners in its creation. It should be updated annually in synchronization with the PPBES process.

• The naval aviation strategic S&T plan should highlight and track naval aviation capability gaps. While information on many of these gaps will come from OPNAV (N70, Warfare Integration, analysis), the Navy Warfare Development Command (NWDC), and the Naval Aviation Strategic Plan, identification of others will be based on external studies and analysis. It is critical that the naval aviation strategic S&T plan be coordinated with the S&T plans of the Air Force, the Army, and NASA. The Joint Aeronautical Commanders Group (JACG) has been a key player in accomplishing this coordination. The Commander, Naval Air Systems Command, is the current chair of the JACG, and the committee understands that he will be reactivating the JACG's coordination of aviation S&T plans among its members.

• The committee recommends that ONR use a systems approach in the planning and development of future naval aviation S&T. The IPPD method used by the Air Force in its S&T planning and implementation is a requirements-driven approach worthy of consideration by ONR.

• The Chief of Naval Research should strengthen ONR's analytic capabilities. A cadre of systems analysis personnel who can interface between the mission capability analysis personnel at OPNAV and NWDC and the scientists and technologists at ONR is needed to support strategic planning for naval aviation S&T.

- With the establishment of a naval aviation strategic S&T plan and the identification of critical gaps in capabilities for naval aviation, ONR should inform and educate congressional staffers about technologies and capabilities that would significantly advance the closure of such gaps, thus turning a currently burdensome relationship into a strategic supportive force.
- The naval aviation strategic S&T plan should track technology progress and investment levels annually for each major and minor technology category. It should track each gap and the investment being made to fill that gap. PEs and associated projects should be tagged for annual analysis in terms of portfolio investment, mission capability gaps, platform categories, and so on.
- The Chief of Naval Research should establish a single point of responsibility for the development of a naval aviation strategic S&T plan at ONR. This responsibility must include both budget and direction-setting authority, even though the technology development will occur in several different organizations. This would enable development of a prioritized, balanced, and well-integrated program that has a high probability of transitioning technology into the operational naval forces.

Appendixes

A

Terms of Reference

At the request of the Office of Naval Research (ONR), the Naval Studies Board (NSB) of the National Research Council will conduct a study to identify promising naval aviation science and technology (S&T) opportunities, to include basic research (6.1), applied research (6.2), and advanced technology development (6.3) areas. Specifically, the study will:

- Review naval and joint operational concepts (e.g., Naval Power 21, Joint Vision 2020) and plans for future naval and joint aviation;
- Identify naval aviation capabilities as a means to employ those operational concepts and plans, including any capability gaps not accounted for or for which no requirements exist; and
- Recommend S&T opportunities (6.1, 6.2, and 6.3) to ONR that could support future naval aviation capabilities and address any capability gaps.

In addition, the study will provide recommendations for naval aviation efforts in the 2007 to 2010 and 2011 to 2025 time frames, and it will categorize capabilities and capability gaps as (1) naval unique, i.e., only required by naval missions; (2) naval essential, i.e., important for naval missions and for non-naval missions as well; and (3) naval relevant, i.e., useful for naval missions and for non-naval missions as well. Finally, while the review will not provide organizational recommendations, it may provide budgetary recommendations in the context of any programs that should be increased, decreased, or eliminated.

B

Committee Meeting Agendas

SEPTEMBER 28-30, 2004

Keck Center of the National Academies
500 Fifth Street NW, Washington, DC 20001

Tuesday, September 28, 2004

Closed Session: Committee Members and NRC Staff Only

0800 CONVENE—Welcome, Composition and Balance Discussion
Dr. Joseph Reagan, Committee Chair
Dr. Charles Draper, NSB Acting Director
Dr. Dennis Chamot, Division on Engineering and Physical Sciences, National Research Council

0900 COMMITTEE DISCUSSION—Motivation for Study, Terms of Reference, Study Plans, Other Topics
Dr. Joseph Reagan, Committee Chair
Dr. Charles Draper, NSB Acting Director
Mr. James Killian, Study Director

Data-Gathering Meeting Not Open to the Public: Classified Discussion (Secret)
SESSION FOCUS: ONR Funding, Scope, Planning, Effectiveness of S&T

1000 OFFICE OF NAVAL RESEARCH—Overview of Organization, Investment Strategy, Planning Process, Effectiveness, Sponsor Goals and Objectives
Dr. Stephen Lubard, Technical Director, ONR 03
Mr. Michael B. Deitchman, Head, Naval Expeditionary Warfare S&T Department, ONR 35

Data-Gathering Meeting Not Open to the Public: Classified Discussion (Secret)
SESSION FOCUS: Naval Power 21, Naval Aviation Operational and Technical Requirements, Global Aviation S&T, ONR Naval Aviation S&T

1145 OFFICE OF THE DEPUTY CHIEF OF NAVAL OPERATIONS FOR PLANS, POLICY, AND OPERATIONS N3/N5—Overview of Naval Power 21 and Sea Strike, Sea Shield, Sea Basing, and FORCEnet
CAPT Thomas E. Mangold, Jr., USN, Head, Strategy and Concepts Branch, Office of Deputy Chief of Naval Operations for Plans, Policy, and Operations, N513

1245 OFFICE OF THE DEPUTY CHIEF OF NAVAL OPERATIONS FOR WARFARE REQUIREMENTS AND PROGRAMS N6/N7—Naval Aviation Operational and Technical Requirements; Sea Strike, Sea Shield, Sea Basing, FORCEnet Capability Gaps Analysis
Mr. Arthur H. Barber, Director, Analysis, Programming, and Integration Division, N70
CDR Danny E. Kowalski, USN, Head, Technology Branch, Office of the Deputy Chief of Naval Operations for Warfare Requirements and Programs, N706

1345 OFFICE OF NAVAL RESEARCH BASIC RESEARCH
Mr. Clifford W. Anderson, Program Officer, ONR 353

1400 OFFICE OF NAVAL RESEARCH—Overview of Integrated High-Performance Turbine Engine Technology/Versatile Affordable Advanced Turbine Engines
Dr. Steven G. Fishman, Program Officer, ONR 332

1445 OFFICE OF NAVAL RESEARCH—Overview of Naval Aviation S&T: Near-, Mid-, Long-Term Capabilities and Investments
Ms. Mun-Won Chang, Aircraft and UAV Technologies Program Manager, ONR 351

Closed Session: Committee Members and NRC Staff Only

1600 COMMITTEE DISCUSSION—Discussion of Day 1, Wrap-up
Moderator: Dr. Joseph Reagan, Committee Chair

1900 END SESSION

Wednesday, September 29, 2004

Closed Session: Committee Members and NRC Staff Only

0830 CONVENE—Committee Discussion, Plans for Day 2
Dr. Joseph Reagan, Committee Chair
Mr. James Killian, Study Director

Data-Gathering Meeting Not Open to the Public: Classified Discussion (Secret)
SESSION FOCUS: Naval Aviation RD&A and Related Issues

0900 OFFICE OF THE SECRETARY OF DEFENSE—Overview of Organization, Aviation Initiatives (e.g., National Aerospace Initiative)
Dr. Spiro Lekoudis, Director, Weapon Systems, Office of the Secretary of Defense (DDR&E)

1000 OFFICE OF THE DEPUTY COMMANDANT FOR AVIATION—Marine Corps Aviation
LtGen Michael A. Hough, USMC, Deputy Commandant for Aviation, Headquarters Marine Corps

1100 PROGRAM EXECUTIVE OFFICER (PEO) TACTICAL AIR PROGRAMS
CAPT Randolph L. Mahr, USN, PEO T

1215 PEO AIR ANTISUBMARINE WARFARE, ASSAULT, AND SPECIAL MISSION PROGRAMS
CDR Richard C. Muldoon, Deputy PEO A for Operations

1315 PEO STRIKE WEAPONS AND UNMANNED AVIATION
Mr. Paul Achille, Naval Air Systems Command
Mr. George M. Clessas, Naval Air Systems Command

1415 NAVAL AIR SYSTEMS COMMAND—Common Systems
Mr. Martin R. Ahmad, Program Manager, NAVAIR

1515 PEO JOINT STRIKE FIGHTER (JSF)
Mr. John C. McKeown, Technical Director, PEO JSF

1545 PEO SPACE AND NAVAL WARFARE SYSTEMS COMMAND (SPAWAR)
CAPT Michael D. Huff, Technical Director, Airborne Networking and Integration IPO, PEO SPAWAR

1615 PEO INTEGRATED WARFARE
Mr. Douglas L. Marker, Advanced Technology Coordinator, PEO Integrated Warfare Systems

Closed Session: Committee Members and NRC Staff Only

1645 COMMITTEE DISCUSSION—Discussion of Day 2, Wrap-up
Moderator: Dr. Joseph Reagan, Committee Chair
1730 END SESSION

Thursday, September 30, 2004

Closed Session: Committee Members and NRC Staff Only

0830 CONVENE—Committee Discussion, Plans for Day 3
Dr. Joseph Reagan, Committee Chair
Mr. James Killian, Study Director

Data-Gathering Meeting Not Open to the Public: Classified Discussion (Secret)
SESSION FOCUS: DOD and Other Aviation Initiatives

0830 NAVAL AIR SYSTEMS COMMAND
Dr. Donald P. McErlean, Deputy Assistant Commander for Research and Engineering, NAVAIR
0930 DEFENSE ADVANCED RESEARCH PROJECTS AGENCY
Dr. Arthur Morrish, Director, Tactical Technology Office, DARPA
1030 U.S. ARMY
Mr. Bruce S. Tenney, Associate Director, Technology, Army Aviation Applied Technology Directorate
1200 U.S. AIR FORCE
Mr. Douglas L. Bowers, Associate Director, Air Platforms, Air Vehicles Directorate, Air Force Research Laboratory
1315 NATIONAL AERONAUTICS AND SPACE ADMINISTRATION
Mr. Robert E. McKinley, Associate Program Manager, Vehicle Systems, Aeronautics Research Mission Directorate, NASA Headquarters

Closed Session: Committee Members and NRC Staff Only

1415 COMMITTEE DISCUSSION—Discussion of Day 3, Plans Ahead, Report Deliberations
Moderator: Dr. Joseph Reagan, Committee Chair
1600 ADJOURN

OCTOBER 26-27, 2004

Keck Center of the National Academies
500 Fifth Street NW, Washington, DC 20001

Tuesday, October 26, 2004

Closed Session: Committee Members and NRC Staff Only

0830 Convene, Welcome Remarks, Review of September Meeting, Study Discussion, Day 1 Plans
Dr. Joseph Reagan, Committee Chair
Mr. James Killian, Study Director

Data-Gathering Meeting Not Open to the Public: Classified Discussion (Secret)

0900 NAVAIR Status of Naval Aviation Strategic Plan, Process
Mr. John D. Robusto, Executive Director, Technology, Intelligence, and Special Programs, Naval Air Systems Command

1030 Chief of Naval Research Overview of ONR and Department of Navy S&T
RADM Jay M. Cohen, USN, Chief of Naval Research

1230 ONR Details on ONR Naval Aviation-Related Projects
CAPT John Hobday, USN, ONR Code 31, Deputy Department Head (30 min.)
Dr. Robert Pohanka, ONR Code 33, Acting Department Head (30 min.)
Dr. Harold Guard, ONR Code 34, Department Head (30 min.)
Ms. Adrienne Gould, ONR Code 36, Deputy Department Head (30 min.)
Dr. John A. Montgomery, NRL Director of Research; CAPT Charles W. Fowler, USN, Interim Military Deputy to the ONR Technical Director (30 min.)

1515 DASN (RDT&E) Review of Department of Navy S&T Including Technology Oversight Group Actions and FNC Alignment Status
Dr. Michael F. McGrath, Deputy Assistant Secretary of the Navy (RDT&E)

Closed Session: Committee Members and NRC Staff Only

1630 Committee Discussion—Recap of Day 1
Dr. Joseph Reagan, Committee Chair
Mr. James Killian, Study Director

1730 End Session

Wednesday, October 27, 2004

Closed Session: Committee Members and NRC Staff Only

0800 Convene, Committee Discussion, Day 2 Plans
Dr. Joseph Reagan, Committee Chair
Mr. James Killian, Study Director

Data-Gathering Meeting Not Open to the Public: Classified Discussion (Secret)

0830 N78 Naval Aviation Operational and Technical Requirements
CAPT Malcolm Taylor, USN (Ret.), S&T Director, Air Warfare Division, Office of the Chief of Naval Operations, N78

0915 DARPA J-UCAS Overview
Mr. John Kinzer, J-UCAS X-47 Deputy Program Manager, DARPA

1030 ONR Naval Aviation S&T Investments
Mr. Michael B. Deitchman, Head, Naval Expeditionary Warfare S&T Department, ONR Code 35

1100 NWDC Results of Naval-Aviation-Related Experimentation/Modeling
Mr. Wayne Perras, Sea Trial Director, Navy Warfare Development Command

Closed Session: Committee Members and NRC Staff Only

1300 Committee Discussion—Meeting Summary, Plans Ahead
Dr. Joseph Reagan, Committee Chair
Mr. James Killian, Study Director

1600 Adjourn

NOVEMBER 16-18, 2004

Arnold and Mabel Beckman Center, Irvine, CA 92612

November 16-18, 2004: Closed Sessions (Committee Members and NRC Staff Only)

0830 CONVENE—PLANS FOR THE DAY
Dr. Joseph B. Reagan, Committee Chair
Mr. James Killian, Study Director

0900 REPORT DISCUSSION AND DRAFTING

1300 CONTINUE REPORT DISCUSSION AND DRAFTING

1700 END SESSION

C

Committee and Staff Biographies

Joseph B. Reagan, Chair, is retired vice president and general manager of research and development at Lockheed Martin Missile and Space and retired vice president and corporate officer of the Lockheed Martin Corporation. Dr. Reagan joined Lockheed as a scientist in 1959, where he led the Space Instrumentation Group for 10 years and was responsible for the development and on-orbit deployment of over 20 scientific payloads for the National Aeronautics and Space Administration and the Department of Defense. His research interests included the areas of space sensors, radiation belt and solar particles, nuclear weapons effects, and the effects of radiation particles on spacecraft systems. Later, as general manager of the research and development division, he led over 750 scientists and engineers in the development of advanced technologies in the fields of optics, electrooptics, information software, guidance and controls, electronics, and materials. Today, he is a director of Southwall Technologies, Inc., a company that manufactures energy-selective thin films for the automotive, electronic, and architectural industries, and a director of SM&A, a leading company in proposal management and program assurance. Dr. Reagan is a member of the National Academy of Engineering and a fellow of the American Institute of Aeronautics and Astronautics and was vice chair of the National Research Council's Naval Studies Board from 2000 to 2004. He received his B.S. and M.S. degrees in physics from Boston College and his Ph.D. in space science from Stanford University.

Frank Alvidrez is senior program manager at Lockheed Martin Company for the Defense Advanced Research Projects Agency Northrop Grumman/Lockheed Martin/Pratt & Whitney X-47B team responsible for a transformational joint unmanned combat air vehicle program. He is a former Marine Corps aviator with expertise in unmanned aerial vehicles, low observable technologies, advanced interoperable enterprise architectures, command and control concepts, advanced systems engineering, network-centric operations and requirements, autonomy technologies, and survivability requirements for unmanned systems. He received his B.S. in petroleum engineering from the University of Southern California and his M.B.A. from National University. He is a graduate of the Defense Acquisition University, Executive Program Management course, and a Department of Defense certified enterprise architect. Mr. Alvidrez has been part of Lockheed Martin Aeronautics Company's Advanced Development Programs for 19 years.

Alfred O. Awani is currently director of advanced tactical lasers at Boeing's Laser and Electro-Optical System Organization. Dr. Awani's expertise is in large-scale systems integration, engineering analysis, design and development, autonomous systems, test and evaluation, technology development and management, systems engineering and requirements development, platform integration, and program management. He has held other key management positions at Boeing and was the Boeing Sikorsky Joint Program Office's deputy director of systems engineering and chief of technology for the Boeing Sikorsky team on the Army Comanche RAH-66 program. Before joining Boeing, he was a research engineer at the National Aeronautics and Space Administration's Ames Research Center, involved in various advanced configuration developments. Prior to his NASA assignment, he was an assistant instructor of aircraft flight dynamics and an instructor of engineering project management at the University of Kansas. He is the recipient of several national and international honors and awards, including the 2002 International Scientist of the Year and the 2001 U.S. Black Engineer of the Year. He earned his B.S. in aeronautical engineering at the Aerospace Institute, a dual M.S. in management and aerospace engineering at Northrop University, and a Ph.D. in engineering from the University of Kansas.

Willard R. Bolton is technical director for the Atmospheric Radiation Measurement-Unmanned Aerospace Vehicles Program, a Department of Energy collaboration involving industrial, academic, and national laboratory participation, and manager of the Exploratory Systems Technology Department at Sandia National Laboratories. Dr. Bolton has an extensive background in aerodynamics, particularly stability and control of unmanned aerial vehicles. Prior to joining Sandia, he was an engineer at the Boeing Military Airplane Division. His professional experience, in both technical areas and program management, has included responsibilities for a number of advanced development and exploratory projects in areas ranging from parachute aerodynamics to high-speed penetration of water, ice, and earth by suborbital missile payloads. He received his B.S., M.S., and Ph.D. in aerospace engineering from the University of Kansas.

William C. Bowes retired from the U.S. Navy with the rank of vice admiral. He is an independent consultant and serves as a director for two public companies and one university board, after having served as vice president of strategic planning at Litton Industries and leading a business unit of Northrop Grumman Electronic Systems. His background is in military operations, naval aviation, systems engineering, life-cycle management, and program management. While in the Navy he served as commander of the Naval Air Systems Command and principal deputy assistant secretary of the Navy for research, development, and acquisition. He is a member of the American Institute of Aeronautics and Astronautics, the Society of Experimental Test Pilots, and the Association for Unmanned Vehicle Systems International. He received a B.S. in chemical engineering from the University of Idaho and an M.S. from the Naval Postgraduate School. He is also a graduate of the U.S. Navy Test Pilot School.

H. Lee Buchanan is executive vice president of Perceptis, LLP, a holding company of wireless and intelligence (government) companies, where he provides expertise in mergers and acquisitions. His experience is in advanced, very high risk, high-payoff technologies such as optoelectronics, laser materials, high-temperature superconductivity, electronic warfare, signal processing, composite structural materials, high-energy-density power sources, and biological warfare defense. Previously, he served as assistant secretary of the Navy for research, development, and acquisition and as deputy director of the Defense Advanced Research Projects Agency. In addition, he was with TITAN Corporation where he founded, built, and directed the company's scientific activities in applied physics, chemistry, and biology. Prior to that he was a senior physicist at Lawrence Livermore National Laboratory, conducting theoretical and experimental research into the physics of directed-energy weapons. Dr. Buchanan also served as a naval flight officer. He earned his B.S. and M.S. in electrical engineering from Vanderbilt University and a Ph.D. in applied physics from the University of California, Berkeley/Davis.

John A. Corder retired from the U.S. Air Force with the rank of major general. His background is in U.S. Air Force operational and joint issues. Since his retirement, he has been employed as an independent technical adviser. He has served as an ad hoc adviser to the Air Force Scientific Advisory Board on the subjects of theater battle management, theater air defense, and tactical ballistic missile defense. General Corder is a former command pilot and navigator. His military career also included assignments as director of electronic combat and commander of the 65th Electronic Combat Air Division with the U.S. Air Force in Europe. He was deputy commander for air combat operations for the Central Command Air Forces in the 1991 Persian Gulf War. General Corder was responsible for the planning and execution of 3,000 combat sorties per day—an effort that involved the coordination of Air Force, Navy, Marine Corps, and allied aircraft from nine other nations. He earned a B.A. in business administration from the University of Nebraska and an M.B.A. from Auburn University.

Robert W. Day is director of space programs at the Raytheon Corporation. His background is in combat command, control, communications, computers, and intelligence systems. Mr. Day joined Raytheon through its merger with the

Hughes Aircraft Company, where he was deputy manager of defense systems. Prior to joining Raytheon, he served in the U.S. Navy for 26 years, during which time he flew A-6 aircraft combat missions. In Washington, Mr. Day served on the staff of the Office of the Chief of Naval Operations as a requirements officer for air warfare and as a division manager for technology requirements. His last duty assignment was as director of stealth and counterstealth technology, where he was responsible for all technology developments, testing, technology transfer, security, export policy, and interservice contacts in the area of stealth and counterstealth. Mr. Day received his B.A. in chemistry from Wesleyan University and an M.B.A. from George Washington University.

Earl H. Dowell is the William Holland Hall Professor and dean emeritus of the Edmund T. Pratt, Jr., School of Engineering at Duke University. A member of the National Academy of Engineering, Dr. Dowell has research interests in aerodynamics, air-breathing propulsion, computational mechanics, energy and power technologies, and structural dynamics. His current research interests include the dynamics of nonlinear fluid and structural systems and their associated limit cycle and chaotic motions. Dr. Dowell has served on numerous scientific boards and advisory committees and currently is a member of the National Research Council's Air Force Studies Board. He received a B.S. from the University of Illinois and his S.M. and Sc.D. from the Massachusetts Institute of Technology.

Valerie J. Gawron is a fellow at General Dynamics. Her experience is in cognitive and environmental effects of human performance, with a specialization in situational awareness, workload, testing, and evaluation as well as engineering psychology and human factors covering the areas of design, research, simulation, and training. She has produced numerous simulation programs, training manuals, and standardized test procedures to improve aviation. She was a member of the Air Force Scientific Advisory Board from 1996 to 2000 and has been a member of the Army Science Board since 2001. Currently, she serves as chair of the Science and Technology Working Group of the National Aeronautics and Space Administration's Space-Human Factors Engineering Group, is a consulting editor of the *International Journal of Aviation Psychology,* and is an associate fellow of the American Institute of Aeronautics and Astronautics and a fellow of the Human Factors and Ergonomics Society. She has over 270 publications. Dr. Gawron earned a B.A. in psychology, an M.S. in industrial engineering, and an M.B.A. in business administration from the State University of New York at Buffalo; an M.A. in psychology from the State University College at Geneseo; and a Ph.D. in engineering psychology from the University of Illinois.

Frank A. Horrigan is retired from the technical development staff for sensors and electronic systems at Raytheon Systems Corporation. He has broad general knowledge of all technologies relevant to military systems. Dr. Horrigan, a theoretical physicist, has more than 40 years of experience in advanced electronics, electrooptics, radar and sensor technologies, and advanced information systems.

In addition, he has extensive experience in planning and managing investments in independent research and development and in projecting future technology growth directions. Dr. Horrigan once served as a NATO fellow at the Saclay Nuclear Research Center in France. Today he serves on numerous scientific boards and advisory committees, including the National Research Council's (NRC's) Army Research Laboratory Technical Assessment Board and the Naval Studies Board. He recently chaired the NRC's Committee for the Review of the Office of Naval Research's Marine Corps Science and Technology Program. He earned a Ph.D. in theoretical physics from Harvard University.

Arun R. Palusamy is currently program manager of systems development and technology at Northrop Grumman Corporation. His expertise is in survivability, countermeasures, and low observables; integration of electronic warfare systems and avionics development; missile defense; sensor/communications design requirements and addressing shipboard integration for next-generation mission systems; and integration of national sensors with naval airborne and surface assets to prosecute boost and midcourse-phase ballistic missile intercept missions. Mr. Palusamy received his B.S. in economics and math from Allegheny College and his M.S. in aeronautical and astronautical engineering from the University of Illinois at Champaign-Urbana.

Robert J. Polutchko is currently manager of strategic business initiatives at the Charles S. Draper Laboratory. He is an experienced engineer with expertise in navigation systems, autonomous vehicles, weapons and aircraft simulations, aircraft avionics, and mission planning systems. Mr. Polutchko was the leader of the precision weapons technology business group responsible for management of the development and demonstration of revolutionary new tactical weapons and system capabilities for U.S. military and industry customers. He has 20 years of research and development experience as a guidance, navigation, and control engineer and manager working on the development of such systems for manned and unmanned spacecraft, collaborative autonomous vehicles, and precision guided weapons systems. Mr. Polutchko earned his B.S. and M.S. degrees in aeronautical engineering from the Massachusetts Institute of Technology.

Bruce Powers is an adjunct professor at George Washington University and a visiting professor at the Naval Postgraduate School. His background is in military requirements, concept development, and naval aviation. He is retired from government service, having spent his last 4 years as the head of strategic assessments and readiness assessments, where he managed mid-range, cost-constrained planning; provided wide-ranging options for the future Navy; and assessed current fleet readiness. He also served as special assistant for plans and analysis—a position in which he was responsible for tying naval aviation to national security policy, strategy, and technology and produced the annual Naval Aviation Plan, a vision for the coming 20 years. He was awarded a second Superior Civilian Service Medal by the Vice Chief of Naval Operations in 2001. Mr. Powers earned an M.S. in physical chemistry from the University of Chicago and an M.S. in industrial engineering/operations from the Illinois Institute of Technology.

Lyle H. Schwartz is currently an independent consultant. He was a professor of materials science and engineering at Northwestern University for 20 years and director of Northwestern's Materials Research Center for 5 of those years. He then became director of the Materials Science and Engineering Laboratory at the National Institute of Standards and Technology, where he served for more than 12 years. His experience there included metals, ceramics, polymers, magnetic materials, techniques for characterization, and standardization of these characterization techniques, and his responsibilities included management of the research and development agenda in the context of a government laboratory. Dr. Schwartz subsquently assumed responsibility for basic research on structural materials of interest to the U.S. Air Force, in addition to the areas of propulsion, aeromechanics, and aerodynamics. He then completed his government service as director of the Air Force Office of Scientific Research, with responsibility for the entire basic research program of the Air Force. His interests include government policy for research and development, particularly for materials research and development and enhanced public understanding of the roles and importance of technology in society. He is a member of the National Academy of Engineering. Dr. Schwartz received a B.S. in engineering and a Ph.D. in materials science from Northwestern University.

William A. Sirignano is engineering dean at the University of California at Irvine. He was a professor at Princeton University from 1967 to 1979 and was the George Tallman Ladd Professor and department head at Carnegie Mellon University from 1979 to 1984. His experience is in combustion, aerospace propulsion, combustion instability, spray and droplet science and technology, noise suppression, and applied mathematics; analysis of driving mechanisms for combustion instability in rockets and ramjets; nonlinear fluid dynamics theory for flame spread above liquid and solid fuels; resolution of turbulent flame and propagation in reciprocating and rotary internal combustion engines; and miniature combustor technology. He is a member of the National Academy of Engineering. Dr. Sirignano received his Ph.D. from Princeton University.

Staff

Charles F. Draper is director of the NRC's Naval Studies Board. Before joining the NRC in 1997, Dr. Draper was the lead mechanical engineer at S.T. Research Corporation, where he provided technical and program management support for satellite Earth station and small satellite design. He received his Ph.D. in mechanical engineering from Vanderbilt University in 1995; his doctoral research was conducted at the Naval Research Laboratory (NRL), where he used an atomic force microscope to measure the nanomechanical properties of thin-film materials. In parallel with his graduate student duties, Dr. Draper was a mechanical engineer with Geo-Centers, Incorporated, working on-site at NRL on the development of an underwater X-ray backscattering tomography system used for the nondestructive evaluation of U.S. Navy sonar domes on surface ships.

James E. Killian is a senior program officer at the National Research Council's National Materials Advisory Board and a retired U.S. Navy captain. During his 26-year career in the Navy he served as the commanding officer of an aircraft carrier-based A-7 Corsair II squadron, commanding officer of the Navy's Nuclear Weapons Evaluation Facility in Albuquerque, New Mexico, and program manager for the Navy's Theater Nuclear Warfare Program (PMS-423) in Washington, D.C. He has a B.S. from the U.S. Naval Academy in Annapolis, Maryland, and an M.S. in aeronautical engineering from the Naval Postgraduate School in Monterey, California.

D

Allocation of Funding in the Naval Aviation Program at the Office of Naval Research

The Deputy Assistant Secretary of the Navy (DASN) for Research, Development, Test, and Evaluation (RDT&E) briefed the committee and provided a breakout of funding allocated to naval aviation technology at the Office of Naval Research (ONR) (Figure D.1). It showed that the President's budget for ONR in fiscal year (FY) 2004 was $1,743 million, with investments in three major categories:

- Discovery and Invention (D&I; categories 6.1 and early 6.2),
- Exploitation and Deployment (E&D; categories late 6.2 and 6.3), and
- Other.

D&I programs were funded at $697.4 million, or 40 percent of the budget. Of this amount, $448.5 million was allocated to internal ONR programs, $157.7 million to the Naval Research Laboratory base of support, and the remaining $91.3 million to various Office of the Secretary of Defense (OSD) programs. E&D funding of $844.5 million, amounting to 48 percent of the budget, included the categories of Future Naval Capabilities (FNCs), Naval Innovations, Marine Corps Warfighting Laboratory programs, and some OSD programs. The FNCs (categories 6.2 and 6.3) were funded at $500.3 million, or 29 percent of the budget. Innovative Naval Prototypes (INPs) activities were initially funded at $225.5 million, but congressional appropriations (plus-ups) of $433 million increased the total to $658.5 million, or 38 percent of the budget. The "Other" category total of $201.3 million (12 percent) includes $164.7 million of pass-

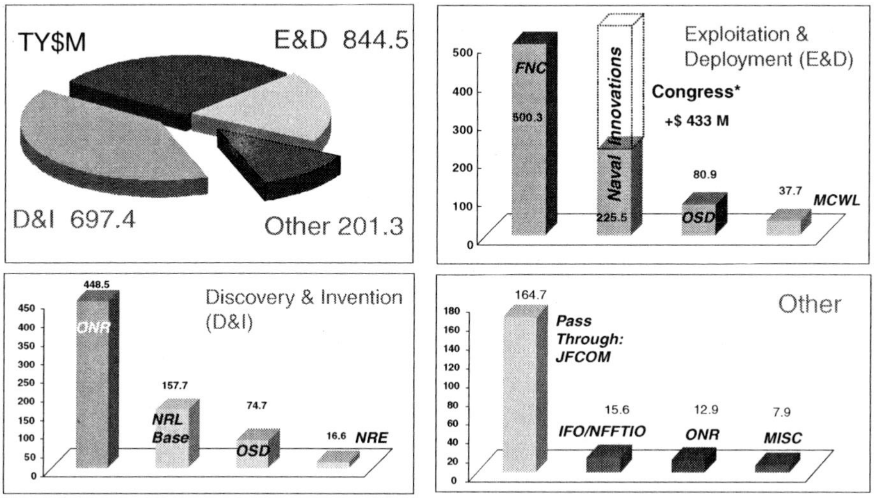

* The FY 04 appropriation added $528 M to Department of Navy S&T, including $433 M for Exploitation and Deployment.

FIGURE D.1 President's FY 2004 Budget for Office of Naval Research Science and Technology. SOURCE: Michael McGrath, Deputy Assistant Secretary of the Navy (RDT&E), "DoN S&T Planning," presentation to the committee on October 26, 2004.

through funding from the U.S. Joint Forces Command and some minor funding from various sources.

In 1999 the Department of the Navy adopted the FNC process as a means of concentrating and focusing its existing 6.2 and 6.3 S&T activities to achieve some near-term technology transitions into the fleet. The Department of the Navy Science and Technology Corporate Board, consisting of the Vice Chief of Naval Operations, the Assistant Commandant of the Marine Corps, and the Assistant Secretary of the Navy for Research, Development, and Acquisition (ASN[RDA]), selects and prioritizes the investments in the FNCs. Each FNC is directed by an integrated product team that brings the perspectives of all the stakeholders to bear: requirements, acquisition, S&T, Office of the Chief of Naval Operations (OPNAV), and operating forces. While notable in the attempt to transition technology to the users, the FNCs are faced with the daunting task of getting the most out of S&T activities that were not planned and optimized from the start to meet user requirements.

INPs are high-risk/high-payoff projects that could be potential "game changers." They exceed FNC limits in both risk and schedule. The criteria for an INP selection were not explained to the committee; however, the DASN RDT&E in his presentation stated that the INPs currently lacked advocacy beyond the Chief

TABLE D.1 Fiscal Year 2004 Funding for Naval Aviation at ONR (millions $)

ONR Dept.	Dept. Name	ONR	ONR + Other
ONR31	Information/electronics	107	160
ONR32	Oceans	9	10
ONR33	Materials/physical	19	35
ONR34	Human systems	4	11
ONR35	Naval expeditionary	218	262
ONR36	Industrial programs	1	10
	TOTAL	358	488

of Naval Research (CNR), and so transition beyond the demonstration phase was doubtful.

While FNCs (near-term benefits) and INPs (potential game changers) make sense in principle, the committee is concerned that only 40 percent of the Navy's S&T budget is being spent on the D&I seed corn of the future. The committee is unaware of any portfolio analysis that showed that this investment distribution was appropriate.

The committee asked for a breakout of the funding dedicated to naval aviation across all ONR departments. A FY 2004 breakout of the 6.1, 6.2, and 6.3 accounts was provided by ONR Code 35 and is shown in Table D.1. Naval aviation received initial funding of $358 million, or 21 percent of the total ONR budget in FY 2004, with Code 35 receiving about 61 percent of this funding.

The naval aviation allocation increases to 28 percent when the Defense Advanced Research Projects Agency and congressionally directed funding is also included. The contributions of the Air Force and Army aviation S&T to naval aviation objectives are not incorporated into these numbers.

Is discretionary annual funding of $358 million and total funding of $488 million, including add-ons amounting to some 21 to 28 percent of the total ONR S&T budget, sufficient for naval aviation? The committee notes that naval aviation consumes 30 to 40 percent of the total Navy procurement budget; therefore, on that basis the 21 to 28 percent share of the S&T budget might seem low. Due to the lack of detailed technology plans, the committee found it difficult to determine the adequacy of current funding based on the information provided. Even if the current S&T funding level is adequate, are the funds optimally allocated across the accounts and ONR codes? The committee concluded that the allocation distribution may not be optimal.

The ONR Code 35 department expressed particular concern over projected significant decreases in its year-to-year upcoming funding. Of special concern to the committee was the minuscule amount of current D&I funding in the 6.1 category performed by this organization, despite the fact that it is the lead code for naval aviation S&T. The committee was told that the 6.1 funding level in FY

2004 was $4.5 million but was scheduled to be reduced to approximately $2.5 million in FY 2005. Clearly there is a problem for the future of naval aviation when the D&I category in the principal naval aviation S&T department is approaching only 1 percent of the budget available to that department and when the total S&T funding for the organization is projected to decrease by over 50 percent over the next few years.

The DASN(RDT&E) briefed the committee and discussed an S&T review for ASN(RD&A), which included the programs and processes at ONR.[1] He concluded that ONR needs a strategic plan for developing S&T and a process for assessing the overall balance of the S&T portfolio. The CNR provided a personal briefing to the committee on the status and achievements of ONR. He acknowledged the current lack of and need for a naval aviation strategic S&T plan and committed his organization to work with the Naval Air Systems Command (NAVAIR) in the creation of one. The Chief Technology Officer of NAVAIR briefed the committee and also committed his organization to work with ONR to develop a strategic S&T plan for naval aviation over the next year. AIR-00X (NAVAIR's S&T manager) acknowledged that there is no formal process for the assessment or transitioning of 6.1 to 6.3 aviation-related efforts at ONR, that naval aviation technology leaders are spread across various laboratories and commands, and that NAVAIR's current strategy is limited to exploiting near-term gains. AIR-00X opined that the aviation visionaries now in industry need to be tapped to bring new perspectives into NAVAIR and ONR. The Deputy Commandant for Aviation, HQMC, briefed his view of the future of Marine Corps aviation and its role in Naval Power 21. OPNAV N70 described the use of war-fighting scenario analyses to identify key capability gaps and to both support the budget process and guide the path of the FNCs.

[1]Michael McGrath, Deputy Assistant Secretary of the Navy (RDT&E), "DoN S&T Planning," presentation to the committee on October 26, 2004.

E

Acronyms and Abbreviations

A/D	analog-to-digital
AFRL	Air Force Research Laboratory
ASN(RDA)	Assistant Secretary of the Navy for Research, Development, and Acquisition
AWS	abrupt wing stall
C4I	command, control, communications, computers, and intelligence
CFD	computational fluid dynamics
CNAF	Commander of Naval Air Forces
CNR	Chief of Naval Research
DARPA	Defense Advanced Research Projects Agency
DASN	Deputy Assistant Secretary of the Navy
DDR&E	Deputy Director of Research and Engineering
D&I	Discovery and Invention
DOD	Department of Defense
E&D	Exploitation and Deployment
FNC	Future Naval Capability
FY	fiscal year
FYDP	Future Years Defense Plan

GPS	Global Positioning System
HPM	high-power microwave
HQMC	Headquarters, U.S. Marine Corps
HSI	human systems integration
IHPTET	Integrated High-Performance Turbine Engine Technology program
IMDS	information management and display system
INP	Innovative Naval Prototype
IP	Internet Protocol
IPDT	integrated product development team
IPPD	Integrated Product and Process Development
IR	infrared
JACG	Joint Aeronautical Commanders Group
J-UCAS	Joint Unmanned Combat Air System
LADAR	laser detection and ranging
LO	low observable
MEMS	microelectromechanical system
MMW	millimeter wave
NASA	National Aeronautics and Space Administration
NAVAIR	Naval Air Systems Command
NCDP	Naval Capabilities Development Process
NRC	National Research Council
NRL	Naval Research Laboratory
NSB	Naval Studies Board
NWDC	Navy Warfare Development Command
ONR	Office of Naval Research
OPNAV	Office of the Chief of Naval Operations
OSD	Office of the Secretary of Defense
PEO	program executive officer
PPBES	Planning, Programming, Budgeting, and Execution System
R&D	research and development
RD&A	research, development, and analysis
RDT&E	Research, Development, Test, and Evaluation
RF	radio frequency

S&T	science and technology
STOM	Ship-to-Objective Maneuver
STOVL	short takeoff and vertical landing
UAV	unmanned aerial vehicle
UCAV	unmanned combat air vehicle
UCAV-N	Unmanned Combat Aerial Vehicle-Navy
USAF	U.S. Air Force
USMC	U.S. Marine Corps
USN	U.S. Navy
VAATE	Versatile, Affordable, Advanced Turbine Engine program
VLO	landing gear operation speed (aviation)
VSTOL	vertical short takeoff and landing